Lecture Notes in Mathematics

Volume 2326

This series reports on new developments in all areas of mathematics and their applications - quickly, informally and at a high level. Mathematical texts analysing new developments in modelling and numerical simulation are welcome. The type of material considered for publication includes:

1. Research monographs
2. Lectures on a new field or presentations of a new angle in a classical field
3. Summer schools and intensive courses on topics of current research.

Texts which are out of print but still in demand may also be considered if they fall within these categories. The timeliness of a manuscript is sometimes more important than its form, which may be preliminary or tentative.

Titles from this series are indexed by Scopus, Web of Science, Mathematical Reviews, and zbMATH.

Steven G. Krantz

The E. M. Stein Lectures on Hardy Spaces

Springer

Steven G. Krantz
Department of Mathematics
Washington University
St. Louis, MO, USA

ISSN 0075-8434   ISSN 1617-9692   (electronic)
Lecture Notes in Mathematics
ISBN 978-3-031-21951-1   ISBN 978-3-031-21952-8   (eBook)
https://doi.org/10.1007/978-3-031-21952-8

This Springer imprint is published by the registered company Springer Nature Switzerland AG
The registered company address is: Gewerbestrasse 11, 6330 Cham, Switzerland

To the memory of E. M. Stein, for his teaching and his friendship.

Preface

Elias M. Stein was one of the pre-eminent harmonic analysts of the twentieth century. He directed more than 50 Ph.D. students at Princeton University, and many of them are quite distinguished mathematicians in their own right. So his influence continues on into the twenty-first century.

One of Stein's seminal contributions to modern mathematical analysis is the real variable theory of Hardy spaces. Hardy spaces were first developed by F. Riesz and M. Riesz in the early part of the twentieth century. It was in fact F. Riesz who, in 1923, named these spaces in honor of G. H. Hardy. These were spaces of holomorphic functions on the unit disc in the complex plane. The Hardy spaces, or H^p, are important because of their structural properties and also because of their behavior under certain important mappings.

People long suspected that, lurking in the background, there is a real-variable theory of Hardy spaces—one that rejects the complex variable context in which they were originally formulated. A real variable theory would allow us to focus on the essential structure and mapping properties of these spaces.

The pioneering work along these lines was done by E. M. Stein and Guido L. Weiss in 1960. There they realized that the right way to formulate a real-variable H^p space was as the gradients of harmonic functions. A number of important calculations and insights appear in their *Acta Mathematica* paper.

A 1971 paper by Burkholder et al. [BGS] provides a glimpse of what is possible in this direction. This paper depends decisively on probabilistic methods.

The paper that broke the subject wide open was the 1972 paper of C. Fefferman and E. M. Stein. This paper is one of the most highly cited in modern mathematics. It inspired a flood of work in the 1970s and on into the 1980s, and continues to have a strong influence today. A major AMS Summer Workshop was held in Williamstown, Massachusetts in 1978 to celebrate and study this paper of Fefferman/Stein and its consequences.

The present book is based on a year-long course that Stein taught at Princeton University in 1973–1974. He in fact taught the course at the request of Robert Fefferman and this author. It was a huge and protracted effort for him to produce this course on the spot, and the results were stunning.

The course that Stein taught was wide-ranging, deep, and insightful. In the characteristic Stein manner, he not only stated theorems and proved them but also worked examples, formulated conjectures, and handed out open problems. Every lecture was an object lesson and a valuable commodity.

This author wrote up very careful notes of the Stein course on Hardy spaces. In fact he wrote up notes *while he was sitting in the course in Fine Hall*. But then, a year or two later, he rewrote and polished the notes a second time. The book that we are presenting here is a formal development of those notes. The purpose now is to share with the mathematical world the perspective of Stein on this subject area that he invented. It is a glowing look back at one of the milestones of modern mathematics. It is a tribute to E. M. Stein.

The content of E. M. Stein's 1973–1974 lecture course, "Real Variable Hardy Spaces," has been reproduced by Steven G. Krantz with permission from the Stein estate.

Of course all errors and mis-steps contained herein are the responsibility of the author. We look forward to hearing from readers as the book is read and appreciated.

St. Louis, Mo, USA   Steven G. Krantz

Contents

Chapter 1
Introductory Material

1.1 Various Maximal Functions

Stein began this first lecture by giving us three references: [RUD1, STW1, ZYG].
   We begin by considering the space

$$(\mathbb{R}_+^2)^n \equiv \mathbb{R}_+ .$$

Thus

$$z \in \mathbb{R}_+ \Longrightarrow z = x + iy , \ x \in \mathbb{R}^n , \ y \in \mathbb{R}^n .$$

The notation $y > 0$ means $y_j > 0$, $j = 1, \ldots, n$. Notice that $\mathbb{R}_+ \subseteq \mathbb{C}^n$ canonically.
   Now $\mathbb{R}_+$ has two boundaries:

 (i) The topological boundary $= \{(x_1 + iy_1, \ldots, x_n + iy_n) : y_j = 0, \ \text{some } j\}$
 (ii) The distinguished boundary $= \{(x_1 + iy_1, \ldots, x_n + iy_n) : y_j = 0 \ \text{for all } j\}$

The distinguished boundary is homeomorphic to $\mathbb{R}^n$ and is equal to the Šilov boundary.

Definition 1.1.1 Let $u(z_1, z_2, \ldots, z_n)$ be a function of n variables, each $z_j \in \mathbb{C}$. Then u is *multiply harmonic* or *n-harmonic* if it is C^∞ and harmonic in each variable.

Lemma 1.1.2 *Suppose that $u : \mathbb{R}_+ \to \mathbb{C}$ is multiply harmonic, bounded, and continuous on $\overline{\mathbb{R}_+}$. Furthermore, assume that $u\big|_{\text{disting. bdry.}} \equiv 0$. Then $u \equiv 0$.*

Proof Suppose without loss of generality that $n = 2$. Fix $z_2 \in \mathbb{R}_+^2$ and consider $u(\,\cdot\,, z_2)$ on $\mathbb{R}_+^2$

$$\sup_{z_1} |u(z_1, z_2)| = \sup_{x_1 \in \mathbb{R}} |u(x_1, z_2)|$$

by the maximum modulus principle.

But for each x_1, $|u(x_1, \,\cdot\,)|$ takes its maximum when z_2 is on $\partial \mathbb{R}_+^2$, that is when $\mathrm{Im} z_2 = 0$. Therefore

$$\sup |u(z_1, z_2)| \leq \sup_{(z_1, z_2) \in \text{dist. bdry. } b\mathbb{R}_+} |u(z_1, z_2)| = 0\,.$$

That completes the proof.   □

Corollary 1.1.3 *We have that*

$$\sup_{z} |u(z)| = \sup_{z \in \text{dist. bdry. } b\mathbb{R}_+} |u(z)|\,.$$

Let us now define the Poisson kernel for $\mathbb{R}_+$ by

$$P_y(x) = P_{y_1}(x_1) \cdot \cdots \cdot P_{y_n}(x_n)\,,$$

where here

$$P_{y_j}(x_j) \equiv \frac{1}{\pi} \cdot \frac{y_j}{x_j^2 + y_j^2}\,.$$

Then P has the following properties:

(1) $\int_{\text{dist. bdry. } \mathbb{R}_+ = \mathbb{R}^n} P_y(x)\,dx = 1$, each $y > 0$. This is proved trivially from the classical case.
(2) Let $U \ni 0$ be any open set. Then

$$\int_{cU} P_y(x),\, dx \to 0 \text{ as } y \to 0\,.$$

This is proved from the classical case or by inspection.
(3) If $f \in L^\infty(\mathbb{R}^n)$ and $u(x, y) \equiv P_y * f$, then u is n-harmonic.

Lemma 1.1.4 *If f is continuous and bounded on the distinguished boundary of $\mathbb{R}_+$ and if*

$$u(x, y) = (\text{multiple}) PI\ f\,,$$

then

(a) *u is multiply harmonic.*
(b) *u is bounded and $\|u\|_{L^\infty} \leq \|f\|_{L^\infty}$.*
(c) *u is continuous on $\overline{\mathbb{R}_+}$.*
(d) $u\Big|_{\text{dist. bdry. } \mathbb{R}_+} = f.$

Proof These are straightforward from **(1)**, **(2)**, and **(3)** above or check the classical case. □

Problem 1.1.5 On $\mathbb{R}_+^2$, u is harmonic if and only if $\Delta u \equiv 0$. This is in turn true if and only if $y^2 \Delta u \equiv 0$. It happens that $y^2 \Delta$ is the only second order differential operator that is invariant under the linear fractional transformations.

The analogue for this operator on $\mathbb{R}_+^2 \times \mathbb{R}_+^2$ is $y_1^2 \Delta_1 + y_2^2 \Delta_2$. This operator degenerates on $\partial \mathbb{R}_+$ and does so strongly on the distinguished boundary.

A theorem of Furstenberg says that if $u : \mathbb{R}_+ \to \mathbb{C}$, u bounded, and $(y_1^2 \Delta_1 + y_2^2 \Delta_2)u = 0$, then $\Delta_1 u \equiv \Delta_2 u \equiv 0$. His proof is rather deep. The problem is to give an elementary proof.

It is worth noting that $\log(y_1/y_2)$ is annihilated by $y_1^2 \Delta_1 + y_2^2 \Delta_2$ but is not multiply harmonic (it is also not bounded).

Definition 1.1.6 Let $f \in L^p(\mathbb{R}^n)$, $1 \leq p \leq \infty$. We define the *strong maximal function*

$$M_S f(x) \equiv \sup_R \frac{1}{m(R)} \int_{y \in R} \int |f(x-y)| \, dy,$$

where R ranges over rectangles of the form

$$R = \{y : |y_j| \leq a_j\}.$$

Let M be the maximal function defined in the same way except that R ranges only over cubes $\{y : |y_j| \leq a\}$. Clearly $Mf(x) \leq M_S f(x)$.

Remark 1.1.7 In the past, one has considered $\mathbb{R}^n$ to be $\partial \mathbb{R}^{n+1}$. We now think of $\mathbb{R}^n$ as $db(R_+^2)^n$ (where db stands for distinguished boundary). The two theories are completely different. In the latter case, many expected things are false.

Recall that

(i) $\|Mf(x)\|_{L^p} \leq A_p \|f\|_{L^p}$,, $1 < p \leq \infty$.
(ii) $m\{x : Mf(x) > \alpha\} \leq (A/\alpha)\|f\|_{L^1}$, $0 < \alpha < \infty$.
    It follows easily from **(ii)** that
(iii) If f is supported on a cube Q_0, then

$$\left| \int_{Q_0} |Mf|^p \, dx \right|^{1/p} \le A \|f\|_{L^1} , 0 < p < 1.$$

Under the hypotheses of **(iii)**, we also have

(iv)

$$\int_{Q_0} |Mf| \, dx \le A \int_{Q_0} |f|(1 + \log^+ |f|) \, dx$$

and more generally

(v)

$$\int_{Q_0} |Mf|(1 + \log^+ Mf)^r \, dx \le A \int |f|(1 + \log^+ |f|)^{r+1} \, dx .$$

For the strong maximal function, one has the following theorem:

Theorem *We have*

(1) $\|M_S f\|_{L^p} \le B_p \|f\|_{L^p}$, $1 < p \le \infty$.

One proves this result by iteration of the estimate for M on $\mathbb{R}^1$. Here $B_p \sim (A_p)^n$. Recall that $A_p \sim 1/(p-1)$ as $p \to 1$.

(2) *There exists $f \in L^1$ such that $M_S f = +\infty$ everywhere.*
(3) *If f is supported in a cube Q_0 and*

$$\int_{Q_0} |f|(1 + \log^+ |f|)^{n-1} \, dx < \infty,$$

then $M_S f < \infty$ a.e. in Q_0 and $M_S f \in L^p(Q_0)$ for $p < 1$. This follows from the iteration argument indicated in (I).
(4) *There exists an f such that $\int f(1 + \log^+ |f|)^{n-1-\epsilon} \, dx < \infty$ but $M_S f = \infty$ everywhere.*

The positive results in this theorem are mainly due to Jessen, Marcinkiewicz, and Zygmund during the period 1930–1934 (cf. *Fundamenta Mathematicae*). Some of the other results are due to Saks.

Corollary 1.1.8 *If $f \in L^p$, $1 < p \le \infty$, or $f \in L(\log L)^{n-1}(Q_0)$, then*

$$\lim_{\text{diam } R \to 0} \frac{1}{m(R)} \int_R f(x - t) \, dt \to f(x) \quad \text{a.e. } x \in Q_0.$$

This follows by standard arguments.

We now indicate how some of the counterexamples indicated in the theorem are obtained.

(2) If a weak type inequality held for all functions, then it would hold for all measures, in particular for the Dirac measure. Now, on $\mathbb{R}^2$,

$$M_S(\delta)(x) = \frac{1}{|x_1||x_2|}.$$

So

$$\{x : M_S(\delta)(x) > \alpha\} = \left\{x : \frac{1}{|x_1||x_2|} > \alpha\right\} = \left\{x : |x_1||x_2| < \frac{1}{\alpha}\right\}.$$

One can easily compute that this region has infinite area.

Remark 1.1.9 Note, by comparison, that $\{x : M\delta(x) > \alpha\} = \left\{x : |x|^2 < \frac{1}{\alpha}\right\}$, which is OK.

Next, we construct a positive, finite measure $d\mu$ which is supported in the unit square and such that $M_S(d\mu) = +\infty$ a.e. (for a reference, see [STE4]). The main lemma is the following, which is a Borel–Cantelli type of result.

Lemma 1.1.10 *Suppose that $E_1, E_2, \ldots$ is an infinite collection of subsets of the unit square Q_0 such that*

$$\sum m(E_i) = \infty.$$

Then there exist sets F_n which are translates of the E_n, each n, so that almost every point of Q_0 is contained in infinitely many F_ns. Equivalently,

$$m\left\{Q_0 \cap \left(\bigcap_{k=1}^{\infty} \bigcup_{j=k}^{\infty} F_j\right)\right\} = m(Q_0) = 1.$$

Now, for $\alpha > 1$, we have

$$m[Q_0 \cap \{x : M_S\delta(x) > \alpha\}] = m\left\{x : |x_1 x_2| < \frac{1}{\alpha}, |x_j| < 1\right\}$$

$$= 2 \cdot \int_0^{1/\alpha} 2\,dx + 4\int_{1/\alpha}^1 \frac{1}{x_1}\frac{1}{\alpha}\,dx$$

$$\approx C\left[\frac{1}{\alpha} + \frac{\log \alpha}{\alpha}\right]$$

$$\approx C\frac{\log \alpha}{\alpha}.$$

Let

$$E_n = \{x \in Q_n : M_S\delta(x) > n\log^2 n\}.$$

Then

$$m(E_n) \sim \frac{\log[n\log^2 n]}{n\log^2 n} \sim \frac{1}{n\log n}.$$

So

$$\sum m(E_n) = +\infty.$$

Let F_n be the sets given by the lemma, each F_n a translate of E_n. Let μ_n be the corresponding translate of δ. Finally, let

$$d\mu = \sum \frac{1}{n(\log n)^{1+\epsilon}} d\mu_n.$$

This is a finite, positive measure. But

$$M_S(d\mu)(x) \geq \frac{1}{n(\log n)^{1+\epsilon}} M_S(d\mu_n)$$

$$> \frac{n\log^2 n}{n(\log n)^{1+\epsilon}}$$

$$= (\log n)^{1-\epsilon}, \text{ each } n, \text{ for } x \in F_n.$$

Since almost every point of Q_0 is in infinitely many of the F_ns, this shows that $M_S(d\mu) = +\infty$ a.e.

1.2 Nontangential Convergence

Let $f \in L^p(\mathbb{R}^N)$ and let $u(x, y)$ be the multiple Poisson integral of f. We define the analogue of the nontangential maximal function for the product setting.

Let $y = (y_1, y_2, \ldots, y_n) \in \mathbb{R}^n$, $y > 0$. Let

$$\Gamma_j = \{z_j \in \mathbb{R}^1 : |x_j| < y_j\}.$$

Set

$$\Gamma = \Gamma_1 \times \cdots \times \Gamma_n = \{z : |x_j| < y_j, j = 1, \ldots, n\}.$$

This is the product cone.
Define

$$M_U f(x) \equiv \sup_{z \in \Gamma(x)} |u(z)|.$$

Lemma 1.2.1 *We have*

(i) $M_U f(x) \leq M_S f(x).$

(ii) *If $f \geq 0$, then*

$$M_S f(x) \leq C \cdot M_U f(x).$$

Proof Suppose withoutloss of generality that $n = 2$. Assume that $f \geq 0$. Then

$$u(x, y) = \frac{1}{\pi^2} \int_{\mathbb{R}^2} \frac{y_1 y_2 f(x_1 - t_1, x_2 - t_2)}{(y_1^2 + t_1^2)(y_2^2 + t_2^2)} \, dt_1 dt_2$$

$$= \sum_{k,j=0}^{\infty} \iint_{\substack{2^{k-1} y_1 \leq |t_1| \leq 2^k y_1 \\ 2^{j-1} y_2 \leq |t_2| \leq 2^j y_2}} + \sum_{j=0}^{\infty} \iint_{\substack{0 \leq |t_1| \leq y_1/2 \\ 2^{j-1} y_2 \leq |t_2| \leq 2^j y_2}}$$

$$+ \sum_{k=0}^{\infty} \iint_{\substack{0 \leq |t_2| \leq y_2/2 \\ 2^{k-1} y_1 \leq |t_1| \leq 2^k y_1}} + \iint_{\substack{0 \leq |t_2| \leq y_2/2 \\ 0 \leq |t_1| \leq y_1/2}}$$

$$\leq \sum_{k,j=0}^{\infty} \iint_{\substack{|t_1| \sim 2^k y_1 \\ |t_2| \sim 2^j y_2}} \frac{1}{y_1 y_2} 2^{-2k} 2^{-2j} f(x_1 - t_1, x_2 - t_2) \, dt_1 dt_2 \qquad \text{plus similar terms}$$

$$\leq \sum_{k,j=0}^{\infty} \iint_{\substack{|t_1| \leq 2^k y_1 \\ |t_2| \leq 2^j y_2}} \frac{1}{(2^k y_1)(2^j y_2)} \cdot 2^{-(j+k)} f(x_1 - t_1, x_2 - t_2) \, dt_1 dt_2.$$

Now fix $x^0 = (x_1^0, x_2^0)$ and suppose that $(x, y) \in \Gamma_{x^0}$. Make the change of variable $t_j \mapsto x_j - x_j^0 + s_j$. Then, since $|x_j - x_j^0| \leq y_j$, we have that the last displayed line is

$$\leq \sum_{k,j=0}^{\infty} \iint_{\substack{|s_1| \leq C 2^k y_1 \\ |s_2| \leq C 2^j y_2}} \frac{1}{(2^k y_1)(2^j y_2)} 2^{-(j+k)} f(x_1 - s_1, x_2 - s_2) \, ds_1 ds_2$$

$$\leq \sum_{k,j=0}^{\infty} 2^{-(j+k)} M_S(f)(x^0)$$

$$\leq C M_S f(x^0).$$

This proves assertion (i). Assertion (ii) is now clear since the $(0, 0)$ term in the sum decomposition of $u(x, y)$ dominates $M_S f$. □

We can define a more general product cone by

$$\Gamma_\alpha(\overline{x}) = \{z : |x_j - \overline{x}_j| < \alpha_j y_j, j = 1, \ldots, n\},$$

where $\alpha = (\alpha_1, \ldots, \alpha_n) > 0$.

Definition 1.2.2 We say that u has an *unrestricted limit* ℓ at $\overline{x}$ if

$$\lim_{\substack{z \in \Gamma_\alpha(\overline{x}) \\ z \to \overline{x}}} u(z) = \ell \quad \text{for all } \alpha > 0.$$

Corollary of the Preceding Lemma *Suppose that* $f \in L^p(\mathbb{R}^n)$ *and that* u *is the multiple Poisson integral of* f. *Then*

(1) *If* $1 < p \le \infty$, *then* u *has unrestricted limits at almost every* $x \in \mathbb{R}^n$ *and the limit function is* f.
(2) *Assertion* (1) *fails in general for* $p = 1$.
(3) *If* f *is continuous and bounded, then* $\lim u(x + iy) = f(x)$ *as* $y \to 0$ *uniformly on compact subsets of* $x \in \mathbb{R}^n$.

Proof Assume that f is bounded and continuous on $\mathbb{R}^n$. Let

$$u(x + iy) = \int P_y(t) f(x - t)\, dt.$$

Hence

$$u(x + iy) - f(x) = \int P_y(t)[f(x - t) - f(x)]\, dt$$

$$= \int_{|t| \le \eta} + \int_{|t| > \eta}$$

$$\equiv T_1 + T_2.$$

Now T_1 may be made $< \epsilon/2$ by choosing η so small that $|f(x - t) - f(x)| < \epsilon/2$ for all $x \in K$ compact—so $|t| < \eta$. Once η is fixed, letting $y \to 0$ makes T_2 vanish. We formalize this last observation: for $\eta > 0$ fixed, $\int_{|t| \ge \eta} P_y(t)\, dt \to 0$ as $y \to 0$ since P is a product of one-dimensional kernels.

Remark 1.2.3 This is about localization. If $f \in L^\infty$ and f vanishes in a neighborhood of $\overline{x}$, then $u(\overline{x} + iy) \to 0$ as $y \to 0$. This follows from the above proof. From this, we can apply results for L^p, $p < \infty$, to an $f \in L^\infty$ by first multiplying f by a $\varphi \in C_c^\infty$. One has $\varphi f \in L^p$ for all $1 \le p \le \infty$.

Example 1.2.4 (Localization Fails in General for Unrestricted Convergence) Precisely, on $\mathbb{R}^2$, with $p < \infty$, there exists $f \in L^p(\mathbb{R}^2)$ with $f \equiv 0$ in a neighborhood of 0, but $u = PI(f)$ does not converge at 0 unrestrictedly.

Fig. 1.1 The function f_1

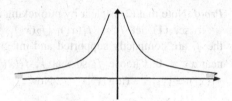

In fact, let $f(x_1, x_2) = f_1(x_1) \cdot f_2(x_2)$, where the f_i are compactly supported, $f_i \geq 0$. Suppose that $f_2(x_2) = 0$ for all x_2 in a neighborhood of 0. We will choose f_1 to be a function that looks like Fig. 1.1, where f_1 blows up at 0 very slowly. Then the multiple Poisson integral of f is $u(x + iy) = u_1(x_1 + iy_1) \cdot u_2(x_2 + iy_2)$ and $u(iy) = u_1(iy_1) \cdot u_2(iy_2)$. For any $y_2 > 0$, $u_2(y_2) > 0$ by the maximum principle and $u_2(y_2) \to 0$ as $y_2 \to 0$. But $u_1(iy_1) \to +\infty$ as $y_1 \to 0$. We need only to choose f_1 so that it blows up at 0 but is locally integrable. Since we are considering *unrestricted convergence*, we let $y_1 \to 0$ so much faster than y_2 that $\lim u_1(y_1) \cdot u_2(y_2) \to \infty$.

Remark 1.2.5 Peaking for the product kernel is on a big subvariety instead of on the diagonal as in $\mathbb{R}^1$. In order for localization to work in the product case, we would have to require that f vanishes in a neighborhood of the whole variety.

Theorem 1.2.6 (Fatou's Theorem) *Suppose that u is multiply harmonic in $(\mathbb{R}_+^2)^n \subseteq \mathbb{C}^n$ and that u is bounded. Then unrestricted limits exist almost everywhere and u is the multiple Poisson integral of some bounded function f. Also, if $1 < p < \infty$, then the condition*

$$\sup_{y>0} \int_{\mathbb{R}^n} |u(x + iy)|^p \, dx < \infty$$

implies that u has unrestricted limits almost everywhere and the limiting function f is in L^p.

Remark 1.2.7 For $p = 1$, these results do not hold in general.

Proof of the Theorem Imitation of the one-dimensional case. □

Definition 1.2.8 Let

$$M_R f(\overline{x}) \equiv \sup_{z \in \Gamma(\overline{x})} |u(z)| \, , \; z_j = x_j + iy_j \, , \; c_1 \leq \frac{y_1}{y_2} \leq c_2 \, , \; k = 1, 2, \ldots, n \, .$$

Lemma 1.2.9 *We have:*

(1) *It is the case that $M_R f(x) \not\leq CMf(x)$. That is to say, there exists a function f so that the left side is ∞ and the right side is $< \infty$ at a point.*

(2) *If $f \geq 0$, then $Mf(x) \leq CM_R f(x)$.*

Proof Note that **(2)** is clear by mimicking the proof of **(ii)** in Lemma 1.2.1.

To see **(1)**, let $f = f_1(x_1) \cdot f_2(x_2)$ with $f_i \geq 0$ for $i = 1, 2$. Assume that the f_i are compactly supported and integrable. Choose f_2 so that $f_2(x_2) = |x_2|$ near $x_2 = 0$. Choose f_1 so that $\int_0^\delta f_1(x_1)\,dx_1 \to 0$ as $\delta \to 0$ very slowly, say $f_1(x_1) = |x_1|^{-1} \cdot (\log |x_1|)^{-3/2}$. Now

$$Mf(0) = \sup_\delta \frac{1}{4\delta^2} \int_{\substack{|x_1| \leq \delta \\ |x_2| \leq \delta}} f_1(x_1) \cdot f_2(x_2)\,dx_1 dx_2 < \infty.$$

But

$$\lim_{\delta \to 0} \iint P_\delta(t_1) \cdot P_\delta(t_2) \cdot f_1(t_1) \cdot f_2(t_2)\,dt_1 dt_2$$

$$\geq \left[\frac{C}{\delta} \int_0^\delta f_1(t_1)\,dt_1 \right] \cdot \left[C\delta \int_\delta^\infty \frac{f(t_2)}{t_2^2}\,dt_2 \right]$$

$$\geq C |\log \delta|^{-1/2} \cdot \left[\int_\delta^\infty \frac{f_2(t_2)}{t_2^2}\,dt_2 \right]$$

$$= C |\log \delta|^{-1/2} \left[\int_\delta^1 \frac{f_2(t_2)}{t_2^2}\,dt_2 + \int_1^\infty \frac{f_2(t_2)}{t_2^2}\,dt_2 \right]$$

$$\geq C |\log \delta|^{-1/2} \cdot |\log \delta|$$

$$= C |\log \delta|^{1/2}$$

$$\to \infty \text{ as } \delta \to 0. \qquad \square$$

Theorem 1.2.10 *Suppose that* $f \in L^p(\mathbb{R}^n)$, $1 \leq p \leq \infty$. *Then* $M_R f(x) < \infty$ *almost everywhere. In fact,*

$$m\{x : M_R f(x) > \alpha\} \leq \frac{A}{\alpha} \|f\|_{L^1}$$

and

$$\|M_R f\|_{L^p} \leq A_p \|f\|_{L^p}, \quad 1 < p \leq \infty, \quad A_p \sim \frac{1}{p-1} \text{ as } p \to 1.$$

If instead $f \in L^1$, *then the restricted limits of the Poisson integral exist almost everywhere and equal* f.

Proof

First Idea: Define, in $\mathbb{R}^2$,

$$R_{k,j} = \{(x_1, x_2) : |x_1| \leq 2^k, |x_2| \leq 2^j\}, \ k = 0, 1, 2, \ldots, \text{ and } j = 0, 1, 2, \ldots.$$

[As usual, we are assuming that $n = 2$.] Let

$$M_{k,j}f(x) = \sup_{y>0} \frac{1}{m(yR_{k,j})} \int_{yR_{k,j}} |f(x-t)| \, dt.$$

Note that $m(R_{k,j}) = 2^{k+1}2^{j+1}$ and $m(yR_{k,j}) = y^2 2^{k+1}2^{j+1}$. It follows that $M_{k,j} = M_{k',j'}$ if $k - j = k' - j'$. It is classical (cf. [STE1]) that

$$m\{x : M_{k,j}f(x) > \alpha\} \leq \frac{C}{\alpha}\|f\|_{L^1}$$

with C independent of k and j. We may let $C = (3)^n = 3^2 = 9$.

Second Idea: If $f_1, f_2, \ldots$ are uniformly of weak type 1, i.e., $m\{x : |f_k(x)| > \alpha\} \leq \frac{1}{\alpha}$, then what can be said about $\sum_{k=1}^{\infty} a_k f_k$? [That is to say, for what choice of a_k is $\sum a_k \frac{1}{|x-x_k|}$ weak type 1 on $\mathbb{R}^1$?] We now give a sufficient condition:

Lemma 1.2.11 *Suppose that* $\sum a_k^{1/2} \leq 1$. *Then* $f = \sum a_k f_k$ *satisfies* $m\{x : |f(x)| > \alpha\} \leq 1/\alpha$ *with* f_k *as above.*

Proof Write $f(x) = \sum a_k f_k(x) = \sum a_k^{1/2}(a_k^{1/2} f_k(x))$. We claim that

$$\{x : |f(x)| > \alpha\} \subseteq \bigcup_{k=1}^{\infty} \{x : a_k^{1/2}|f_k(x)| > \alpha\}.$$

For if $x \notin \bigcup_{k=1}^{\infty}\{x : a_k^{1/2}|f_k(x)| > \alpha\}$, then $a_k^{1/2}|f_k(x)| \leq \alpha$ for all k, so $|f(x)| \leq \sum a_k^{1/2}(a_k^{1/2} f_k(x)) \leq \alpha$. This proves the inclusion. Thus

$$m\{x : |f(x)| > \alpha\} \leq \sum_k m\{x : a_k^{1/2}|f_k(x)| > \alpha\} \leq \frac{\sum a_k^{1/2}}{\alpha} \leq \frac{1}{\alpha}.$$

This completes the proof. □

It should be noted that the best possible condition on the a_k is $\sum |a_k| \log |1/a_k| \leq 1$.

Proof of Theorem 1.2.10 By the lemma and the First Idea, it will suffice to show that

$$MRf(x) \leq C \sum_{k,j=0}^{\infty} 2^{-k-j} M_{k,j} f(x).$$

Let $\overline{x} = (\overline{x}_1, \overline{x}_2)$. Let $z = (z_1, z_2)$ be the variable point, $z_j = x_j + iy_j$. So we suppose that $|x_j - \overline{x}_j| < y_j$. Recall that $P_y(t) = P_{y_1}(t_1) \cdot P_{y_2}(t_2)$ with $P_{y_j}(t_j) = (1/\pi) y_j / (t_j^2 + y_j^2)$.

There are two key estimates:

(1) Since $c_1 \leq y_1/y_2 \leq c_2$, we have $P_{y_1}(\cdot) \sim P_{y_2}(\cdot)$. This is clear.
(2) $P_{y_1}(t_1) \sim P_{y_1}(t_1 + h)$ provided that $|h| \leq c y_1$.

Let us prove this last result.

Suppose without loss of generality that $C > 1$. If $|t_1| \leq 2C |y_1|$, then

$$\frac{1}{\pi} \frac{y_1}{t_1^2 + y_1^2} \leq \frac{1}{\pi} \frac{y_1}{y_1^2} \leq \frac{1}{\pi} \frac{32C^2 y_1}{32C^2 y_1^2} = \frac{1}{\pi} \frac{32C^2 y_1}{16C^2 y_1^2 + 16C^2 y_1^2}. \qquad (*)$$

But

$$|t_1 + h|^2 \leq (3C|y_1|)^2 \leq 16C^2 |y_1|^2$$

so that

$$(*) \leq \frac{1}{\pi} \frac{32C^2 y_1}{(t_1 + h)^2 + 16C^2 y_1^2} \leq \frac{1}{\pi} \frac{32C^2 y_1}{(t_1 + h)^2 + y_1^2}.$$

If $|t_1| \geq 2C|y_1|$, then $|h| \leq |t_1|/2$, and hence

$$\frac{1}{\pi} \frac{y_1}{t_1^2 + y_1^2} \leq \frac{1}{\pi} \frac{y_1}{|t_1 + h|^2 + y_1^2} \leq \frac{1}{\pi} \frac{2y_1}{|t_1 + h|^2 + y_1^2}.$$

Now

$$|u(x_1 + iy_1, x_2 + iy_2)| = \left| \iint P_{y_1}(t_1) P_{y_2}(t_2) f(x_1 - t_1, x_2 - t_2) \, dt_1 dt_2 \right|.$$

Using estimates **(1)** and **(2)** above, we majorize this last by

$$\leq \sum_{\substack{2^{k-1} y \leq t_1 \leq 2^k y \\ 2^{j-1} y \leq t_2 \leq 2^j y}} P_y(t_1) P_y(t_2) |f(\overline{x}_1 - t_1, \overline{x}_2 - t_2)| \, dt_1 dt_2 \quad + \quad \text{other leftover terms}$$

$$\leq \sum_{k,j} y^{-2} \cdot 2^{-2k-2j} \int_{y R_{kj}} |f(\overline{x}_1 - t_1, \overline{x}_2 - t_2)| \, dt_1 dt_2$$

$$\leq C \sum 2^{-k-j} M_{k,j}(f)(\overline{x})$$

as desired. □

Problem 1.2.12 Is there a Fatou theorem for restricted convergence? More precisely, if u is multiply harmonic in $(\mathbb{R}^2_+)^n$ and bounded in the set where $c_1 \leq y_1/2 \leq c_2$, then does u have boundary values in the restricted sense?

We know that the global analogue for Theorem 1.2.6 is true.

Remark 1.2.13 Let $\varphi \geq 0$, $\varphi \in L^1(\mathbb{R}^n)$, $\int \varphi \, dx = 1$. Let $\varphi_\epsilon(x) = \epsilon^{-n}\varphi(x/\epsilon)$. So $\int_{\mathbb{R}^n} \varphi_\epsilon(x) \, dx = 1$ for all $\epsilon > 0$. We ask the following question:

Is $\sup_{\epsilon>0}(|f| * \varphi_\epsilon)$ of weak type 1 when $f \in L^1$?

A sufficient condition is the following: define $\Phi(x) \equiv \sup_{|x'|\geq|x|} \phi(x')$. If $\int_{\mathbb{R}^n} \Phi(x) \, dx < \infty$, then the answer is "yes," for then $\sup_{\epsilon>0} |f| * \varphi_\epsilon \leq AMf$, where $A = \int \Phi$.

Here is the difficulty: the φ which arises from restricted convergence is, in $\mathbb{R}^2$,

$$\varphi(x_1, x_2) = \frac{1}{\pi^2} \frac{1}{1+x_1^2} \frac{1}{1+x_2^2}.$$

Hence

$$\Phi(x_1, x_2) \geq \frac{1}{\pi^2} \cdot 1 \cdot \frac{1}{1+|x|^2},$$

which is not integrable at ∞ in $\mathbb{R}^2$.

One can ask what is the right maximal function which dominates restricted convergence? Here is a possibility:

Let $\Omega : [-\pi, \pi] \to \mathbb{R}^+$. Then one can prove that the operator

$$M_\Omega(f)(x) \equiv \sup_{h>0} \frac{1}{h^2} \int_0^h \int_{-\pi}^{\pi} |f(x + re^{i\theta})| \Omega(\theta) r \, d\theta \, dr$$

satisfies

$$\|M_\Omega f\|_{L^p} \leq A_p \|f\|_{L^p} \|\Omega\|_{L^1(\text{circle})}, \quad 1 < p \leq \infty.$$

One uses the method of rotations on the one-dimensional result (see [STW1]). Here, as usual, $A_p = 1/(p-1)$ as $p \to 1$.

An interesting special case of these ideas is when Ω is a characteristic function. Then M_Ω amounts to a generalized Hardy–Littlewood maximal function taken over star-shaped balls. The ball is a union of radii through E where $\Omega = \chi_E$; see Fig. 1.2.

The point of this discussion is

$$M_R f(x) \leq A_\epsilon M_{\Omega_\epsilon}(f)(x), \qquad (*)$$

Fig. 1.2 A picture of the ball

where $\Omega_\epsilon(\theta) = |\sin 2\theta|^{-2\epsilon}$ and $A_\epsilon \sim 1/\epsilon$ as $\epsilon \to 0$. We now prove inequality $(*)$.

Observation If $\mu(x) \geq 0$ is radial and decreasing as a function of $r = |x|$ on $\mathbb{R}^2$, if $\int_{\mathbb{R}^2} \mu(x)\,dx < \infty$, and if we let $\varphi(r, \theta) = \mu(r) \cdot \Omega(\theta)$, then

$$\sup_{\epsilon > 0}(|f| * \varphi_\epsilon)(x) \leq A\|\mu\|_{L^1} M_\Omega(f)(x). \qquad (\star)$$

This is immediate when $\mu(r) = \alpha\chi_{[0,r_0]}(r)$, some $r_0 > 0$.

The general result follows by taking monotone limits of finite positive linear combinations of such functions.

By the above remarks and the proof of Theorem 1.2.10, it suffices for us to show that if

$$\varphi(x) = \frac{1}{1 + x_1^2} \cdot \frac{1}{1 + x_2^2},$$

then

$$\varphi(r, \theta) \leq \frac{1}{(1 + r^2)^{1+\epsilon}} A|\sin 2\theta|^{-2\epsilon}.$$

For

$$\int_{\mathbb{R}^2} \frac{dx}{(1 + |x|^2)^{1+\epsilon}} \leq \int_{|x| \leq 2} + \int_{|x| > 2}$$

$$\leq C + C \int_{|x| > 2} \frac{dx}{|x|^{2+2\epsilon}}$$

$$\leq \frac{C}{\epsilon}.$$

By $(\star)$, this is enough.

Now

$$\frac{1}{(1 + x_1^2)(1 + x_2^2)} = \frac{1}{1 + r^2 + r^4(\sin\theta\cos\theta)^2}.$$

Since we wish to show that this is

$$\leq \frac{A|\sin 2\theta|^{-2\epsilon}}{(1+r^2)^{1+\epsilon}},$$

we may as well suppose that $r > 1$, otherwise the result is trivial.

So it suffices to show that

$$\frac{(1+r^2)^{1+\epsilon}}{1+r^2+r^4|\sin 2\theta|^2} \leq A|\sin 2\theta|^{-2\epsilon}$$

or, since $r > 1$,

$$\frac{(r^2)^{1+\epsilon}}{r^2+r^4(\sin 2\theta)^2} \leq A|\sin 2\theta|^{-2\epsilon},$$

or, letting $r^2 = s$,

$$\frac{s^{1+\epsilon}}{s+s^2(\sin 2\theta)^2} \leq A|\sin 2\theta|^{-2\epsilon}.$$

With $t = s/(\sin 2\theta)$, this becomes essentially

$$\frac{t^{1+\epsilon}}{t+C^2t^2} \leq C^{-2\epsilon}.$$

But if $t \leq 1/C^2$, then we have

$$\frac{t^{1+\epsilon}}{t+C^2t^2} \leq \frac{t^{1+\epsilon}}{t} \leq C^{-2\epsilon}.$$

If instead $t > 1/C^2$, then

$$\frac{t^{1+\epsilon}}{t+C^2t^2} \leq \frac{t^{1+\epsilon}}{C^2t^2} \leq \frac{1}{C^2}t^{\epsilon-1} \leq C^{-2\epsilon}.$$

That does it.

Remark 1.2.14 Note that, by the method of rotations or by inspection, one can check that M_{Ω_ϵ} is of type (p, p), $1 < p \leq \infty$. The L^1 result, unfortunately, does not follow in this fashion. But at least now we have a geometrical maximal function which dominates restricted convergence.

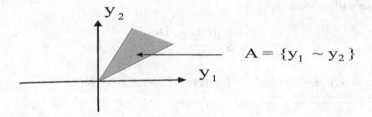

Fig. 1.3 Restricted convergence

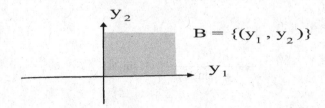

Fig. 1.4 Unrestricted convergence

1.3   Unrestricted Convergence

Now we discuss unrestricted convergence. We remark that when we consider holomorphic functions on $\mathbb{R}_+^2$, instead of just multiply harmonic functions, then the notions of restricted convergence and unrestricted convergence come closer together. Consider $\mathbb{R}_+^2 \times \mathbb{R}_+^2$. The restricted convergence situation is shown in Fig. 1.3.

The unrestricted convergence situation is shown in Fig. 1.4.

Now A can be mapped by a real linear transformation to B. This map can be extended to a complex linear map on $\mathbb{C}^n$ in the obvious way. Call the extended map L. If f is holomorphic, then so is $f \circ L$. This fact enables one to pass from one theory to the other. But, if f is harmonic (or even multiply harmonic), then the same need not be true for $f \circ L$.

Definition 1.3.1 If u is multiply harmonic on $\mathbb{R}_+$ and $x \in \mathbb{R}^n$, then u is said to be unrestrictedly bounded at x provided that there exist an $\alpha > 0$ and $h > 0$ such that u is bounded on $\Gamma_\alpha^h \equiv \Gamma_{\alpha_1}^{h_1}(x_1) \times \Gamma_{\alpha_2}^{h_2}(x_2) \times \cdots \times \Gamma_{\alpha_n}^{h_n}(x_n)$. [Note that there is no relationship among the y_js, hence the name.]

Theorem 1.3.2 Let u be multiply harmonic in $\mathbb{R}_+A$. Let $E \subseteq \mathbb{R}^n$. If u is unrestrictedly bounded at each $x \in E$, then u has unrestricted limits at almost every point of E.

Proof The proof of this result for $\mathbb{R}_+^n$ is well known, so we only indicate in detail where this proof differs from that one.

We begin, as usual, by uniformizing the situation: namely, we must check that it suffices to assume that for some $\alpha > 0$ and $h > 0$,

$$|u(x, y)| \leq 1 \quad \text{for all } (x, y) \in \bigcup_{x_0 \in E} \Gamma_\alpha^h(x_0) \, , \ E \text{ compact.}$$

The proof of this reduction relies on a point-of-density argument. However, in the present situation we must use the fact that almost every point of a set $E \subseteq \mathbb{R}^n$ is a point of *strong density* of E. That is to say,

$$\lim_{\text{diam } R \to 0} \frac{m(E \cap R)}{m(R)} = 1 \, ,$$

where R is any rectangle centered at x with sides parallel to the axes.

After uniformizing things, we suppose that, for $E \subseteq \mathbb{R}^2$ compact, $|u| \leq 1$ in $\mathcal{R} \equiv \cup_{x_0 \in E} \Gamma_\alpha^1(x_0)$. For each $n \geq 1$, set

$$\varphi_n(x_1, x_2) = \begin{cases} u(x_1 + i/n, x_2 + i/n) & \text{if } (x_1 + i/n, x_2 + i/n) \in \mathcal{R} \\ 0 & \text{otherwise.} \end{cases}$$

Let $\varphi_n(z) = $ multiple $\text{PI}[\varphi_n(x)]$. Then, of course,

$$|\varphi_n(x)| \leq 1 \quad \text{for } x \in \mathbb{R}^2 \, , \ n \geq 1.$$

Define $u_n(z) = u(x_1 + i(y_1 + 1/n), x_2 + i(y_2 + 1/n))$, and finally let $\psi_n(z) = u_n(z) - \varphi_n(z)$. All this notation is analogous to the proof for $\mathbb{R}_+^n$.

Now since $\|\varphi_n\|_{L^\infty} \leq 1$, there is a subsequence φ_{n_k} converging weak-$*$ to some $\varphi \in L^\infty(\mathbb{R}^2)$. Let $\varphi(z) = $ multiple $\text{PI}(\varphi)$. Then of course $u_{n_k}(z) \to u(z)$ and $\varphi_{n_k}(z) \to \varphi(z)$, so $\psi_{n_k}(z) \to u(z) - \varphi(z) \equiv \psi(z)$. Since φ is a multiple Poisson integral of an L^∞ function, φ has unrestricted limits almost everywhere. So, in order to see that u has unrestricted limits almost everywhere, it behooves us to show that $\psi(z) \to 0$ as $z \to x_0$ through $\Gamma_\alpha^1(x_0)$ for almost every $x_0 \in E$. As usual, in order to show this, it is enough to find a multiply harmonic $h \geq 0$ in $\mathbb{R}^+$ such that, in $\mathcal{R}$, $|\psi(z)| \leq h(z)$ and h has unrestricted limits 0 almost everywhere on E.

For the function h, we use $h(z) = C\left[y_1 + y_2 + MP[\chi_{cE}]\right]$. Clearly h is multiply harmonic, $h \geq 0$, and h has unrestricted limits 0 almost everywhere in E. We shall check that $|\psi_n(z)| \leq h(z)$ in $\mathcal{R}$, finishing the proof. Of course, we must choose C appropriately.

Now we need only to check our inequality on $b\mathcal{R}$. More precisely, we must show that if $\overline{z} \in b\mathcal{R}$, then

$$\psi_n(\overline{z}) \leq \liminf_{\substack{z' \to \overline{z} \\ z' \in \mathcal{R}}} h(z') \, . \tag{$\star$}$$

Fig. 1.5 The set E

Fig. 1.6 The fact that
$\Lambda \subseteq {}^c E$

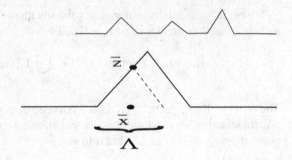

Consider Fig. 1.5.

The boundary of $\mathcal{R}$ consists of four parts:

(1) Either y_1 or y_2 equals h.
(2) $0 < y_1 < h, 0 < y_2 < h$.
(3) $y_1 = 0$ but $0 < y_2 < h$ or vice versa.
(4) $y_1 = 0$ and $y_2 = 0$.

So we wish to check $(\star)$ in each of these four cases.

For **(1)**, just make C sufficiently large.

For **(4)**, note that when $y_1 = y_2$, we have $\psi_n = 0$.

To check **(2)** is more difficult. Let $\overline{z} = (\overline{x}_1 + i\overline{y}_1, \overline{x}_2 + i\overline{y}_2) \in b\mathcal{R}$. We must check that $|\psi_n(\overline{z})| \le h(\overline{z})$.

Now let $\Lambda = \{(x_1, x_2) : |x_j - \overline{x}_j| < \alpha\overline{y}_j\}$. We claim that $\Lambda \subseteq {}^c E$. This is clear pictorially—see Fig. 1.6.

Thus

$$h(\overline{z}) \ge MPI(\chi_\Lambda)$$

$$= \frac{\overline{y}_1\overline{y}_2}{\pi^2} \iint \frac{\chi_\Lambda(\overline{x}_1 - t_1, \overline{x}_2 - t_2)}{(t_1^2 + \overline{y}_1^2)(t_2^2 + \overline{y}_2^2)} \, dt_1 dt_2$$

$$\ge \frac{C}{\overline{y}_1\overline{y}_2} \iint \chi_\Lambda(\overline{x}_1 - t_1, \overline{x}_2 - t_2) \, dt_1 dt_2$$

$$= \frac{C}{\overline{y}_1\overline{y}_2} m(\Lambda)$$

$$\ge C \cdot \psi_n(\overline{z}) .$$

That proves **(4)**.

For **(3)**, let $\overline{z} = (\overline{x}_1, \overline{x}_2 + i\overline{y}_2), 0 < \overline{y}_2$. Set $z' = (x_1' + iy_1', x_2', +iy_2')$. Then

$$MPI(\chi_{c_E})(z_1', z_2') = \frac{y_2'}{\pi^2} \int \frac{dt_2}{t_2^2 + y_2'^2} \cdot y_1' \cdot \int \frac{\chi_{c_E}(x_1' - t_1, x_2' - t_2)}{t_1^2 + y_1'^2} \, dt_1,$$

so

$$\liminf_{y_1' \to 0} MPI(\chi_{c_E})(\overline{x}_1 + iy_1', \overline{z}_2) \geq \frac{1}{\pi^2} \overline{y}_2 \int \frac{\chi_{c_E}(\overline{x}_1, \overline{x}_2 - t_2)}{t_2^2 + \overline{y}_2^2} dt_2, \qquad (*)$$

by Fatou's lemma, once we check that

$$\liminf_{y_1' \to 0} y_1' \int \frac{\chi_{c_E}(\overline{x}_1 - t_1, \overline{x}_2 - t_2)}{t_1^2 + y_1'^2} dt_1 \geq \chi_{c_E}(\overline{x}_1, \overline{x}_2 - t_2).$$

Now in case $(\overline{x}_1, \overline{x}_2 - t_2) \in E$, then the right side is 0 so the inequality is clear. But if $(\overline{x}_1, \overline{x}_2 - t_2) \in {}^cE$, we have, since ${}^cE$ is open, that $\chi_{c_E}(\overline{x}_1 - t_1, \overline{x}_2 - t_2) = 1$ for t_1 small. So the approximate identity property of the Poisson kernel shows that the left side is 1 (as is the right). Thus we have established $(*)$. But it is easy to see that the right side of $(*)$ is bounded from 0, independent of $\overline{z}$, so we have established $(\star)$, hence the theorem. $\square$

Assertion: Somewhere in the above theory, the "distinguished boundary" is playing a role. As usual assume $n = 2$. Let $\overline{z}$ lie in the topological boundary of $\mathcal{R}$. Furthermore, let $\overline{z} = (\overline{z}_1, \overline{z}_2) = (\overline{x}_1 + i\overline{y}_1, \overline{x}_2 + i\overline{y}_2)$. We say that $\overline{z}$ is a *boundary point of type 1* if $(\overline{x}_1 + i\overline{y}_1 - \epsilon, \overline{x}_2 + i\overline{y}_2) \notin \mathcal{R}$ for all sufficiently small $\epsilon > 0$. Similarly, $\overline{z}$ is a boundary point of type 2 if $(\overline{x}_1 + i\overline{y}_1, \overline{x}_2 + i\overline{y}_2 - \epsilon) \notin \mathcal{R}$ for all sufficiently small $\epsilon > 0$. Every nontrivial boundary point is of either type 1 or type 2 or both (a trivial boundary point is one for which $\overline{y}_1 = \overline{y}_2 = 0$ or $\overline{y}_1 = \overline{y}_2 = h$). Let $\Pi = \{\overline{z} \in b\mathcal{R} : \overline{z} \text{ is of type 1 and of type 2}\}$. We conjecture that Π should play the role of the distinguished boundary for this local theory. It is *unknown* whether this can be made to work.

1.4 The Area Integral

Let $n = 2$ and $u(z_1, z_2)$ be a multiply harmonic function in $(\mathbb{R}_+^2)^2$. Let

$$|\nabla_j f|^2 = \left|\frac{\partial f}{\partial x_j}\right|^2 + \left|\frac{\partial f}{\partial y_j}\right|^2, j = 1, 2.$$

Define

$$S(u) = \left(\iint_{\Gamma(x)} |\nabla_1 \nabla_2 u(z)| \, dx dy\right)^{1/2},$$

where $\Gamma(x) = \Gamma_1(x_1) \times \Gamma_2(x_2)$.

One can check that if $u = MPI(f)$, $f \in L^2(\mathbb{R}^2)$, then

$$\int_{\mathbb{R}^2} (S(u))^2 \, dx_1 dx_2 = 4 \int_{\mathbb{R}^2} |f|^2 \, dx_1 dx_2 \, .$$

To prove this, we can use Green's theorem or we can recall that

$$\triangle_1(u^2) = 2|\nabla_1 u|^2$$

and

$$\triangle_1 \triangle_2 (u^2) = 4|\nabla_1 \nabla_2 u|^2 \, .$$

Problem 1.4.1 Show that unrestricted convergence for a multiply harmonic function is locally almost everywhere equivalent with the finiteness of $S(u)$.

Definition 1.4.2 For $1 \le p < \infty$, let

$$H^p(\mathcal{R}_+) = \{ f \text{ holomorphic on } \mathcal{R}_+ : \sup_{y>0} \int_{\mathcal{R}_+} |f(x+iy)|^p \, dx \equiv \|f\|_{H^p}^p < \infty \} \, .$$

[Use the usual variant when $p = \infty$.]

Theorem 1.4.3 *Suppose that $F \in H^p(\mathcal{R}_+)$, $p > 0$. Then*

(1) *$F(x + iy)$ has unrestricted limits as $y \to 0$ for almost every $x \in \mathbb{R}^n$.*
(2) *If $F^*(x) \equiv \sup_{z \in \Gamma(x)} |F(z)|$, then $\|F^*\|_{L^p} \le C\|F\|_{H^p}$.*
(3) *If $p < \infty$ and if $F(x) = \lim_{y \to 0} F(x + iy)$, then*

$$\int_{\mathbb{R}^n} |F(x + iy) - F(x)|^p \, dx \to 0 \text{ as } y \to 0 \text{ (unrestrictedly).}$$

Proof (In Case $n = 2$) First observe that if $F(z_1, z_2)$ is in $H^p(\mathcal{R}_+)$, then $F(\cdot, z_2) \in H^p \cdot$), each fixed z_2. Now fix a z_2 and a $y_1 > 0$. We claim that

$$\int_{-\infty}^{\infty} |F(x_1 + iy_1, z_2)|^p \, dx_1 \le C y_2 \|F\|_{H^p(\mathcal{R}_+)}^p \, .$$

To see this, note that by the subharmonicity of $|F|^p$ in each variable,

$$|F(z_1, z_2)|^p \le \frac{C}{(y_1)^2 (y_2)^2} \iint_{B(z_1, y_1/2)} \iint_{B(z_2, y_2/2)} |F(x_1' + iy_1', x_2' + iy_2')|^p \, dx_1' dy_1' dx_2' dy_2' \, .$$

Thus

$$\int_{-\infty}^{\infty} |F(x_1 + iy_1, z_2)|^p \, dx_1$$

$$\leq \frac{C}{y_1(y_2)^2} \iint_{\substack{x_1' \in \mathbb{R} \\ y_1/2 \leq y_1' \leq 3y_1/2}} \iint_{\substack{x_2' \in \mathbb{R} \\ y_2/2 \leq y_2' \leq 3y_2/2}} |F(x_1' + iy_1', x_2' + iy_2')|^p \, dx_1' dy_1' dx_2' dy_2'$$

$$\leq \frac{C}{y_2} \|F\|_{H^p}^p .$$

This proves the claim.

Now if $F \in H^p(\mathbb{R}_+)$ and z_2 is fixed, then $|F(z_1 + i\epsilon, z_2)|^{p/2}$ is subharmonic in z_1 and has bounded L^2 norms along the lines $y_1 = $ constant, as we just verified. We have in addition

$$|F(z_1 + i\epsilon, z_2 + i\epsilon)|^{p/2} \leq \int_{-\infty}^{\infty} P_{y_1}(x_1 - t_1)|F(t_1 + i\epsilon, z_2 + i\epsilon)|^{p/2} \, dt_1 \qquad \textbf{(a)}$$

by subharmonicity. A similar notion applied to $|F(t_1 + i\epsilon, \cdot)|^{p/2}$ shows that

$$|F(t_1 + i\epsilon, z_2 + i\epsilon)|^{p/2} \leq \int_{-\infty}^{\infty} P_{y_2}(x_2 - t_2)|F(t_1 + i\epsilon, t_2 + i\epsilon)|^{p/2} \, dt_2 . \qquad \textbf{(b)}$$

Putting together **(a)** and **(b)** gives

$$|F(z_1+i\epsilon, z_2+i\epsilon)|^{p/2} \leq \iint_{\mathbb{R}^2} P_{y_1}(x_1-t_1) P_{y_2}(x_2-t_2)|F(t_1+i\epsilon, t_2+i\epsilon)|^{p/2} \, dt_1 dt_2 .$$

$$(*)$$

Now the functions $|F(t_1 + i\epsilon, t_2 + i\epsilon)|^{p/2}$ are uniformly bounded in L^2, so there exists a weak-$*$ convergent subsequence $|F(t_1+i\epsilon_k, t_2+i\epsilon_k)|^{p/2}$ with limit F. Also $\|f\|_{L^2} \leq \|F\|_{H^p}^{p/2}$. Taking the limit of $(*)$ as $k \to \infty$ yields

$$|F|^{p/2}(z) \leq MPI(f)(z)$$

whence

$$(F^*)^{p/2}(x) \leq \sup_{z \in \Gamma(x)} |MPI(f)(z)| \leq M_s f(x)$$

by Lemma 1.2.1. Therefore

$$\|F^*\|_p^p \leq C_p \|f\|_p^p \leq C_p \|F\|_{H^p}^p , \quad p > 0.$$

[Note that we have arranged everything above to take place in L^2 and H^2 so that all relevant operators are bounded.]

Fig. 1.7 The cone Γ

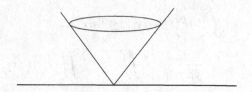

Part (2) is thus proved. Part (1) now follows from the local Fatou theorem. Part (3) follows from (1) and dominated convergence. □

1.5 Generalizations of $\mathbb{R}_+$ and $H^p(\mathbb{R}_+)$

We now pass to generalizations of $\mathcal{R}_+$ and $H^p(\mathcal{R}_+)$ (Fig. 1.7).

Definition 1.5.1 A set $\Gamma \subseteq \mathbb{R}^n$ is called an *open cone* if

(1) $x \in \Gamma, y \in \Gamma \Rightarrow x + y \in \Gamma$.
(2) $x \in \Gamma, r > 0 \Rightarrow rx \in \Gamma$.
(3) $0 \notin \Gamma$.
(4) Γ is open.

Definition 1.5.2 The *tube domain* T_Γ over Γ is $\{z \in \mathbb{C}^n : z = x + iy, y \in \Gamma\}$.

Definition 1.5.3 The cone Γ is said to be *regular* if Γ contains no lines.

Example 1.5.4 In $\mathbb{R}^1$, the only cones are $\{x : x > 0\}$ and $\{x : x < 0\}$.
In $\mathbb{R}^2$, all cones are essentially of the form

$$\{z : 0 < \arg z < \theta\}, \ \theta \leq \pi.$$

The cone is regular if $\theta < \pi$.
In $\mathbb{R}^3$, there are a great many different cones. Two examples are:

(a) The first octant $= \{(y_1, y_2, y_3) : y_j > 0\}$
(b) The forward light cone $= \{(y_1, y_2, y_3) : y_3 > 0, y_1^2 + y_2^2 < y_3^2\}$

Examples (a) and (b) are "homogeneous" in the sense that, given two points of the cone, there exists a linear transformation under which the cone is invariant and one point is moved to the other.
An example of a non-homogenous cone in $\mathbb{R}^3$ is readily seen to be the open, convex hull of four rays emanating from 0 and in general position.

Definition 1.5.5 Let $\Gamma \subseteq \mathbb{R}^n$ be a cone. The *dual cone* Γ^* is defined to be $\Gamma^* \equiv \{y \in \mathbb{R}^n : y \cdot x \geq 0, x \in \Gamma\}$.

It is a fact (see [STW1]) that Γ^* contains an interior point if and only if Γ is regular.

Definition 1.5.6 Define

$$H^p(T_\Gamma) = \{F : F \text{ is holomorphic on } T_\Gamma \text{ and } \sup_{y \in \Gamma} \int_{\mathbb{R}^n} |F(x+iy)|^p \, dx = \|F\|_{H^p}^p < \infty\}.$$

The main theorem on $H^2(T_\Gamma)$ is then the following.

Theorem 1.5.7 *Let Γ be a regular cone. Then if $f \in L^2(\Gamma^*)$, we see that*

$$F(z) = \int_{\Gamma^*} f(t)e^{2\pi i t \cdot z} \, dt$$

converges absolutely to a function which is in $H^2(T_\Gamma)$. Every $F \in H^2(T_\Gamma)$ arises in this fashion from a unique f and the correspondence $f \leftrightarrow F$ is norm preserving, i.e., $\|F\|_{H^2(T_\Gamma)} = \|f\|_{L^2(\Gamma^*)}$.

More precisely, one can say that the integral converges absolutely and uniformly when y remains in a compact subset of Γ. Also

$$\sum_{y \in \Gamma} \int |F(x+iy)|^2 \, dx = \|f\|_{L^2(\Gamma^*)}.$$

It is sometimes convenient to think of a partial ordering given by Γ as follows:

We say that $y > 0$ if and only if $y \in \Gamma$ and $y_1 > y_2$ if and only if $y_1 - y_2 \in \Gamma$.

Remark 1.5.8 Suppose that y_0 is a fixed point in Γ. Then there exists a $\delta_{y_0} > 0$ such that $e^{-2\pi y_0 \cdot t} \le e^{-(\delta_{y_0})|t|}$ whenever $t \in \Gamma^*$.

Proof of the Theorem If y' is *any* unit vector in the y space, not necessarily in the cone, then $y_0 - \epsilon y' \in \Gamma$, ϵ sufficiently small. So

$$(y_0 - \epsilon y') \cdot t \ge 0 \text{ for all } t \in \Gamma^*,$$

i.e.,

$$y_0 \cdot t \ge +\epsilon y' \cdot t.$$

By symmetry,

$$y_0 \cdot t \ge |\epsilon y' \cdot t| \text{ for all unit vectors } y'$$

or

$$y_0 \cdot t \ge \epsilon|t|.$$

This gives the result.

Now clearly

$$\int_{\Gamma*} e^{-2\pi y \cdot t} \, dt \le \int_{\Gamma*} e^{-2\pi \delta_y |t|} \, dt < \infty$$

for any $y \in \Gamma$. By definition, moreover,

$$F(z) = \int_{\Gamma*} e^{2\pi i x \cdot t} e^{-2\pi y \cdot t} f(t) \, dt \, .$$

By Schwarz's inequality,

$$|F(z)| \le \left(\int_{\Gamma*} e^{-4\pi y \cdot t} dt \right)^{1/2} \cdot \|f\|_{L^2(\Gamma*)} < \infty \, .$$

To prove the assertion about uniform convergence, it suffices to consider convex hulls of finitely many points $y^1, y^2, y^3, \ldots, y^N$.

Suppose that y is in the convex hull of $\{y^j\}_{j=1}^N$. So $y = \sum_{j=1}^N a_j y^j$, $\sum a_j = 1$, $a_j \ge 0$. We consider $e^{2\pi i z \cdot t} f(t)$, $\operatorname{Im} z = y$. Then

$$\left| e^{2\pi i z \cdot t} f(t) \right| = e^{-2\pi y \cdot t} |f(t)|$$

$$= e^{-2\pi \sum a_j y^j \cdot t} |f(t)|$$

$$\le \sum_{j=1}^N a_j e^{-2\pi y^j \cdot t} |f(t)| \, .$$

The convergence of $\int_{\Gamma*}$ on the left side hangs, of course, on that of the right side. Since every compact subset of Γ is contained in a finite union of such balls, we are done.

Now, by Plancherel's theorem,

$$\int_{\mathbb{R}^n} |F(x + iy)|^2 \, dx = \int_{\Gamma*} e^{-4\pi y \cdot t} |f(t)|^2 \, dt \quad \text{for } y \in \Gamma$$

$$\le \int_{\Gamma*} |f(t)|^2 \, dt \, .$$

Since

$$\lim_{y \to 0} \int_{\Gamma*} e^{-4\pi y \cdot t} |f(t)|^2 \, dt = \int_{\Gamma*} |f(t)|^2 \, dt \, ,$$

the asserted isometry is proved.

To see that every $F \in H^2(T_\Gamma)$ arises as the integral of an $f \in L^2(\Gamma^*)$, we need a lemma:

Lemma 1.5.9 *Suppose that* $F \in H^2(T_\Gamma)$. *If* K *is relatively compact in* Γ, *then* F *is bounded on* $\mathbb{R}^n \times K$.

Proof Fix K relatively compact in Γ. Find an $\epsilon > 0$ so small that if $y \in K$ and $|y - \overline{y}| < \epsilon$, then $\overline{y} \in \Gamma$. If $z = x + iy$, $y \in K$, let $P(z) = \{w \in \mathbb{C}^n : |z_j - w_j| < \epsilon\}$. Thus

$$|F(z)|^2 \le \frac{1}{(\pi \epsilon^2)^n} \int_{P(z)} |F(z')|^2 \, dx'dy' \le C_K \|F\|^2_{H^2(T_\Gamma)}.$$

Now suppose that $F \in H^2(T_\Gamma)$, $|F(x + iy)| \le A_y(1 + |x|)^{-n-1}$, where the constant A_y is uniform for y in compact subsets of Γ. Then $y \in \Gamma$ implies that

$$\int_{\mathbb{R}^n} F(x + iy)e^{-2\pi ix \cdot t} \, dx \equiv f_y(t)$$

is well defined. We assert that $f_y(t) = e^{-2\pi y \cdot t} f(t)$ gives a well-defined $f(t)$ which meets our requirements.

In order to see that this well defines f, it suffices to have

$$f_y(t) \cdot e^{2\pi y \cdot t} = f_{\overline{y}}(t)e^{2\pi \overline{y} \cdot t} \quad, \text{ all } y, \overline{y} \in \Gamma.$$

Say that $\overline{y} = (\overline{y}_1, \overline{y}_2, \dots, \overline{y}_n)$ and $y = (y_1, y_2, \dots, y_n)$.

We will now use Cauchy's integral formula one variable at a time. Fix $z_2, z_3, \dots, z_n$ and look at

$$F(\cdot, z_2, z_3, \dots, z_n)e^{-2\pi iz \cdot t}.$$

We wish to see that

$$\int F(x_1 + iy_1, z_2, \dots, z_n)e^{-2\pi iz \cdot t} \, dx_1 = \int F(x_1 + i\overline{y}_1, z_2, \dots, z_n)e^{-2\pi i\tilde{z} \cdot t} \, dx_1,$$

where $z = (z_1, z_2, \dots, z_n)$ and $\tilde{z} = (x_1 + i\overline{y}_1, z_2, \dots, z_n)$.

Use the Cauchy Integral theorem on the indicated box (see Fig. 1.8) to verify this inequality. Now repeat this procedure in each variable. This defines $f(t)$. Moreover, Plancherel's formula shows that F and f have the desired relation.

Now we may remove the growth restriction on F; this is where we use Lemma 1.5.9. Look at

$$\left| e^{-\epsilon(z_1^2 + \cdots + z_n^2)} \right| = e^{-\epsilon \sum x_j^2 + \epsilon \sum y_j^2}.$$

Fig. 1.8 The Cauchy integral over a box in the complex plane

If y is contained in a fixed compact set, then this is a rapidly decreasing function of x.

Let $F_\epsilon(z) = e^{-\epsilon \sum_{j=1}^{n} z_j^2} F(z)$ and $f_y(t) = \int F(x+iy)e^{-2\pi ix\cdot t}\, dx$. If, in this last expression, we replace F by F_ϵ and use the first part of the proof, we get

$$(f_\epsilon)_y(t) = \int F_\epsilon(x+iy)e^{2\pi ix\cdot t}\, dx,$$

and the expression on the right is independent of y. So we define

$$f_\epsilon(t) = e^{2\pi y\cdot t}(f_\epsilon)_y(t) = e^{2\pi y\cdot t}\int F_\epsilon(x+iy)e^{2\pi ix\cdot t}\, dx.$$

If we fix y (so that $F(\cdot + iy)$ is bounded by the lemma), then we may let $\epsilon \to 0$ to obtain the desired f. Let us summarize what we have accomplished.

Given $F \in H^2(T_\Gamma)$, we may write

$$F(x+iy) = \int e^{2\pi ix\cdot t}e^{-2\pi y\cdot t}f(t)\, dt,$$

where $e^{-2\pi y\cdot t}f(t)$ is in L^2. We wish to see that f is supported in T^*. But

$$\sup_{y>0}\int_{\mathbb{R}^n}|F(x+iy)|^2\, dx < \infty$$

if and only if

$$\sup_{y\in\Gamma}\int_{\mathbb{R}^n} e^{-4\pi y\cdot t}|f(t)|^2\, dt < \infty.$$

Now if f has support not in Γ, then $-4\pi y \cdot t$ will be positive there and the expression $e^{-4\pi y\cdot t}$ can get very large. This will cause a contradiction. More precisely, let $\bar{t} \notin \Gamma^*$ and let $y_0 \in \Gamma$ satisfy $y_0 \cdot \bar{t} < 0$. There exists an $\epsilon > 0$ such that if $t \in B(\bar{t}, \epsilon)$, then $y_0 \cdot t \leq -\delta < 0$. Now let $y = Ny_0$, where $N > 0$ is large. Then

$$\sup_{y \in \Gamma} \int_{\mathbb{R}^n} e^{-4\pi y \cdot t} |f(t)|^2 \, dt < \infty \Rightarrow \sup_N \int_{B(\bar{t},\epsilon)} e^{-4\pi N y_0 \cdot t} |f(t)|^2 \, dt < \infty$$

$$\Rightarrow \sup_N \int_{B(\bar{t},\epsilon)} e^{4\pi N \delta} |f(t)|^2 \, dt < \infty$$

$$\Rightarrow f(t) \equiv 0 \text{ on } B.$$

So f is supported in Γ^*.

This completes our rather long and chatty proof of Theorem 1.5.7.   □

Corollary 1.5.10 *If $F \in H^2(T_\Gamma)$, then $\lim_{y \to 0} F(x+iy) = f(x)$ exists in L^2 norm and*

$$\|F\|_{L^2} = \|F\|_{H^2(T_\Gamma)} = \|f\|_{L^2(\Gamma^*)} \, .$$

Proof Use Plancherel's theorem.   □

Now we define a Cauchy kernel for T_Γ.

Definition 1.5.11 With Γ as usual, let

$$C(z) = \int_{\Gamma^*} e^{2\pi i z \cdot t} \, dt \ , \ z = x + iy \ , \ y \in \Gamma.$$

This function C will be our Cauchy kernel.

This definition is due to S. Bochner.

Observe that $C(z)$ is holomorphic on T_Γ. We have, with the notation $C_y(x) = C(x + iy)$, $y \in \Gamma$, that

$$C_y \in L^2(\mathbb{R}^n) \bigcap L^\infty(\mathbb{R}^n),$$

and the norms are uniform when y is restricted to compact subsets of Γ.

Proposition 1.5.12 *If $F \in H^2(T_\Gamma)$, then*

$$F(x + iy) = \int C_y(t) F(x - t) \, dt \, .$$

If y_1 and y_2 are in Γ, then

$$F(x + iy_1 + iy_2) = \int_{\mathbb{R}^n} C_{y_1}(t) F(x - t + iy_2) \, dt \, .$$

Proof This is clear from what we have already done. For example, let $\Gamma = \{y \in \mathbb{R}^1 : y > 0\}$. Consider $T_\Gamma = R_+^2$. Then Proposition 1.5.12 is well known.   □

In many cases we can explicitly compute C.

We define the *Poisson kernel* associated with the tube domain T_Γ by

$$P(x, y) = \frac{|C_y(x)|^2}{C(2iy)} , \quad y \in \Gamma.$$

In many cases, P satisfies an appropriate second order differential equation, but in many other cases it does not. This definition of the Poisson kernel is due to Hua.

From the properties of C, we see that

$$P_y \in L^1 \cap L^\infty, \quad \text{each } y \in \Gamma.$$

Also, by Plancherel,

$$\int_{\mathbb{R}^n} |K(x + iy)|^2 \, dx = K(2iy), \tag{$*$}$$

so $P(x, y) \geq 0$.

Theorem 1.5.13 *The Poisson kernel P satisfies the following properties:*

(1) $\int_{\mathbb{R}^n} P_y(x) \, dx = 1$.

(2) *If U is any open set containing 0, then*

$$\lim_{\substack{y \to 0 \\ y \in \Gamma}} \int_U P_y(x) \, dx = 1.$$

(3) *If $F \in H^2(T_\Gamma)$, then*

$$F(x + iy) = \int_{\mathbb{R}^n} P_y(t) F(x - t) \, dt = \int_{\mathbb{R}^n} P_y(x - t) F(t) \, dt.$$

(3') *If $y_1, y_2 \in \Gamma$, then*

$$F(x + iy_1 + iy_2) = \int_{\mathbb{R}^n} P_{y_1}(t) F(x - t + iy_2) \, dt.$$

It is not generally the case that $P_{y_1} * P_{y_2} = P_{y_1 + y_2}$. It is worthwhile investigating in what sense this semigroup property holds.

Proof of the Theorem Part **(1)** is clear from observation $(*)$.

For part **(2)**, we construct an $F \in H^2(T_\Gamma)$ such that F is continuous on $\overline{T_\Gamma}$, $F(0) = 1$, $|F(x)| < 1$ if $x \neq 0$, and $|F(x)| \to 0$ as $|x| \to +\infty$. This is a "peak function" for T_Γ. The existence of such an F, plus the statement **(3)** of the theorem, will yield the result. For pick $\epsilon > 0$. Choose $\delta > 0$ such that $|F(x)| < 1 - \epsilon$ if $|x| > \delta$ and also such that $B(0, \delta) \subseteq U$. Then

$$1 = \lim_{\substack{y \in \Gamma \\ y \to 0}} \left\{ \int_{|x| \le \delta} F(x)\mathcal{P}(x, y)\, dx + \int_{|x| > \delta} F(x)\mathcal{P}(x, y)\, dx \right\} .$$

Thus

$$1 \le \lim_{\substack{y \in \Gamma \\ y \to 0}} \left\{ \int_{|x| \le \delta} \mathcal{P}(x, y)\, dx + (1 - \epsilon) \int_{|x| > \delta} \mathcal{P}(x, y)\, dx \right\}$$

$$\le \lim_{\substack{y \in \Gamma \\ y \to 0}} \left\{ 1 - \epsilon \int_{|x| > \delta} \mathcal{P}(x, y)\, dx \right\} .$$

Therefore

$$\lim_{\substack{y \in \Gamma \\ y \to 0}} \int_{|x| > \delta} \mathcal{P}(x, y)\, dx = 0 .$$

In conclusion,

$$1 = \lim_{\substack{y \in \Gamma \\ y \to 0}} \int_{\mathbb{R}^n} \mathcal{P}(x, y)\, dx = \lim_{\substack{y \in \Gamma \\ y \to 0}} \int_U \mathcal{P}(x, y)\, dx$$

as desired. So we must construct such an F.

In fact we let

$$F(x + iy) = \int_{\Gamma^*} e^{2\pi i z \cdot t} f(t)\, dt ,$$

with $f(t) \ge 0$, smooth, compactly supported in Γ^*, and such that $\int f(t)\, dt = 1$. Then all the assertions we have made about F are obvious except that $|F| < 1$ when $x \ne 0$.

But

$$F(x) = \int_{\Gamma^*} e^{2\pi i x \cdot t} f(t)\, dt .$$

This would be 1 if we were to take the absolute value of the integrand, but as it stands there will be cancellation. This proves (2).

It now behooves us to prove (3) and (3') *without using* (2). Let $F \in H^2(T_\Gamma)$. Fix $w \in T_\Gamma$, $w = u + iv$, $v \in \Gamma$, $u \in \mathbb{R}^n$, and let

$$G(z) = F(z)C(z - \overline{w}) .$$

From our observations about C, it follows that $G \in H^2$. So, by the Cauchy formula,

$$G(w) = \int_{\mathbb{R}^n} C(u - t + iv) F(t) C(t - u + iv) \, dt \, .$$

But, from the definition of C, $\overline{C(x + iy)} = C(-x + iy)$. So

$$F(w)C(2iw) = G(w) = \int_{\mathbb{R}^n} |C(u - t + iv)|^2 F(t) \, dt$$

or

$$F(w) = \int_{\mathbb{R}^n} \frac{|C(u - t + iv)|^2}{C(2iv)} F(t) \, dt = \int_{\mathbb{R}^n} \mathcal{P}_v(t) F(x - t) \, dt \, .$$

The proof of **(3')** is similar, using **(3)**. $\square$

Example 1.5.14 We actually present six examples, some in detail.

1. Let Γ = first octant in $\mathbb{R}^n = \{y : y > 0\}$. Write

$$\mathcal{P}_y(x) = \mathcal{P}_{y_1}(x_1) \cdot \mathcal{P}_{y_2}(x_2) \cdot \cdots \cdot \mathcal{P}_{y_n}(x_n) \, .$$

2. The *Lorentz inner product* is

$$(x, y) = x_1 y_1 - x_2 y_2 - x_3 y_3 - \cdots x_n y_n \, .$$

So

$$(y, y) = y_1^2 - \sum_{j=2}^{n} y_j^2 \, .$$

Let

$$\Gamma = \{y : (y, y) > 0, \, y_1 > 0\} \, .$$

This space is only interesting when $n \geq 3$. For, when $n = 2$, Γ is simply a rotation of the first quadrant. We claim that $\Gamma^* = \overline{\Gamma}$.

Proof of Claim Suppose that $x \in \mathbb{R}^n$ with $x \cdot y \geq 0$ for all $y \in \Gamma$. Then $x \cdot (1, 0, \ldots, 0) \geq 0$, so $x_1 \geq 0$. Suppose that y_j, $j = 2, 3, \ldots, n$, are real numbers satisfying $y_2^2 + y_3^2 + \cdots + y_n^2 = 1$. Then

$$(1, y_2, \ldots, y_n) \in \overline{\Gamma}$$

so that

$$x \cdot (1, y_2, \ldots, y_n) \geq 0,$$

that is,

$$x_1 - \sum_{j=2}^{n} x_j y_j \geq 0,$$

that is,

$$x_1 \geq \sum_{j=2}^{n} x_j y_j.$$

Since the y_j were arbitrary, it follows that

$$\left(\sum_{j=2}^{n} x_j^2\right)^{1/2} \leq x_1$$

or

$$\sum_{j=2}^{n} x_j^2 \leq x_1^2.$$

That is to say, $x \in \overline{\Gamma}$. So $\Gamma^* \subseteq \Gamma$.

The reverse inclusion is trivial.

If we define

$$(z, w) = z_1\overline{w}_1 - z_2\overline{w}_2 - \cdots z_n\overline{w}_n \ , \ z, w \in \mathbb{C}^n,$$

then one can compute that

$$C(z) = c_n(z, z)^{-n/2}, \tag{$*$}$$

where $c_n = \Gamma(n/2)/(2\pi^{n/2}i^n)$. So

$$\mathcal{P}_y(x) = \frac{|c_n|(y, y)^{n/2}}{|(x + iy, x + iy)|^n} = \frac{|c_n|(y, y)^{n/2}}{[((x, x) - (y, y))^2 + 4(x, y)]^{n/2}}.$$

When n is odd, there are subtleties lurking in this formula for C, but the function does not vanish on the cone, so we may define $(z, z)^{-n/2}$ by choosing a branch.

We remark that one computes $(*)$ by computing C at a point and then using the fact that Γ is a homogeneous space under the Lorentz group, where one can define the Lorentz group to be those linear transformations which respect $(\cdot, \cdot)$. See [STE2], where this is carried out in a computation of the Bergman and Cauchy–Szegő kernels.

For example, one can explicitly compute

$$C(i, 0, \dots, 0)) \quad = \quad \int_{\Gamma^*} e^{2\pi i \cdot i \cdot t} \, dt$$

$$= \quad \int_{\Gamma} e^{-2\pi t} \, dt$$

$$= \quad \int_0^\infty m\{(t_2, \dots, t_n) : \sum_{j=2}^n t_j^2 \le t_1^2\} e^{-2\pi t_1} \, dt_1$$

$$= \quad \int_0^\infty V_{n-1} t_1^{n-1} e^{-2\pi t_1} \, dt_1$$

$$= \cdots = V_{n-1}(n-1)!(2\pi)^{-n+1} \int_0^\infty e^{-2\pi t_1} \, dt_1$$

$$= \quad V_{n-1}(n-1)!(2\pi)^{-n},$$

where V_{n-1} is the volume of the unit ball in $(n-1)$-space.

So the forward light cone has the following interesting properties:

(a) Self-duality
(b) Homogeneity
(c) Positivity under a Hermitian inner product

3. Now let $n = [m(m+1)]/2$, $m > 0$, $m \in \mathbb{Z}$. We may identify $\mathbb{R}^n$ with all $m \times m$ real symmetric matrices. Let $S_m \subseteq E^m$ be all positive definite real symmetric matrices. The automorphism group of S_m can be obtained from the general linear group on E_m. Namely, if $g \in GL(E_m)$, $x \in S_m$, let $\rho_g x = gx^t g$. The tube over this cone is the "Siegel generalized upper half plane." Note that in case $m = n = 1$, then $T_{S_m} = \mathbb{R}_+^2$.

One can compute

$$\mathcal{P}_y(x) = \text{constant} \cdot \left| \frac{\det y}{|\det(x+iy)|^2} \right|^{(n+1)/2}.$$

4. When $n = m^2$, we obtain a complex analogue of 3. Namely, we identify $\mathbb{R}^n$ with complex Hermitian matrices and $\mathbb{C}^n$ with $m \times m$ complex matrices. Then, of course, the cone H_m is the set of all $m \times m$ positive definite hermitian matrices. As in 3, the automorphism group on H_m can be obtained from $GL(n, \mathbb{C})$. Namely, if $g \in GL(n, \mathbb{C})$, define $\rho_g x = gx^t \overline{g}$ for $x \in \mathbb{R}^n$.

5. When $n = 2m^2 - m$, one identifies $\mathbb{R}^n$ with the $m \times m$ quaternionic hermitian matrices and carries out the above program.

6. There is an analogue of 3, 4, and 5 for the Cayley numbers in case $n = 16$.

Theorem 1.5.15 (Vinberg, Koecher, and Rothaus) *Every homogeneous self-dual cone is a product of the cases* **1–6**.

One has a good chance of doing some function theory in the above setting, as one can compute the Cauchy and Poisson integrals explicitly, and one can use the symmetries and group actions. There exists a theory of "Jordan algebras," which encompasses the above examples.

1.6   Relationships Among Domains

We now make some remarks on the relationships among the domains we have studied.

Every tube domain over a regular cone is biholomorphically equivalent to a bounded domain in $\mathbb{C}^n$. One sees this as follows: after a linear transformation, surround the cone by the first octant. The first octant is of course a product of half-planes which is biholomorphically equivalent to a product of discs. However, there exist bounded domains that are not biholomorphically equivalent to tubes over cones (e.g., the unit ball). Of course, we now have

bounded domains $\underset{\neq}{\supseteq}$ tube domains over regular cones

$\supseteq$ tube domains over homogeneous regular cones

(there are uncountably many non-equivalent such

domains—cf. Piatetski-Shapiro, Vinberg, etc.)

$\supseteq$ tubes over homogeneous, self-dual cones

(there are finitely many of these in each dimension)

The latter collection coincides with the bounded symmetric domains of Cartan with distinguished boundary having $1/2$ the real dimension of the domain.

Recall that $\mathcal{D} \subseteq \mathbb{C}^n$ is a bounded, symmetric domain of Cartan if

(i) $\mathcal{D}$ is relatively compact in $\mathbb{C}^n$.
(ii) $\mathcal{D}$ is homogeneous in the sense that the group of holomorphic self-mappings of $\mathcal{D}$ is transitive.
(iii) (symmetry) Given $p \in \mathcal{D}$, there exists a mapping $T : \mathcal{D} \to \mathcal{D}$ holomorphic such that $T^2 = \text{id}$ and $T(p) = p$ for p an isolated fixed point.

Examples of bounded symmetric domains of Cartan:

- The unit ball in $\mathbb{C}^n$. If $p = 0 \in B$, then let $T : z \mapsto -z$. Obtain T for any other point in the unit ball by conjugation by an appropriate biholomorphic map.

- On the Siegel domain, the transformation T for the point $(i, 0, \ldots, 0)$ is $T : z \mapsto -1/z$.

It was a longstanding open problem whether the defining conditions (i) and (ii) for a bounded symmetric domain of Cartan implied condition (iii). It turns out that the answer is "no."

Also, it has been shown that tube domains over homogeneous cones do not coincide with tube domains over homogeneous, self-dual cones.

We return now to tube domains over cones. Let Γ be a regular cone. We say that Γ_0 is a proper subcone of Γ if $\overline{\Gamma_0} \subseteq \Gamma \cup \{0\}$.

Definition 1.6.1 If $F : T_\Gamma \to \mathbb{C}, \overline{x} \in \mathbb{R}^n$, we say that the limit of F exists as $z \to \overline{x}$ *restrictedly* if

$$\lim_{\substack{z \to \overline{x} \\ z = x + iy}} F(z)$$

exists, where $|x - \overline{x}| \leq C|y|$, $y \in \Gamma_0$, some Γ_0 a regular subcone of Γ, some $C > 0$.

Theorem 1.6.2 *Let Γ be a regular, convex cone, $F \in H^p(T_\Gamma)$, $0 < p \leq \infty$. Then*

(1) $\lim_{z \to x} F(z) = F(x)$ *holds almost everywhere in the restricted sense.*
(2) $\int_{\mathbb{R}^n} |F(x + iy)| - F(x)|^p \, dx \to 0$ *as $y \to 0$ restrictedly, $0 < p < \infty$.*
(3) *For $p \geq 1$, convergence works in the sense of (2) without restricting y to a subcone.*

Unsolved Problems

(a) What happens to almost everywhere unrestricted convergence?
(b) What happens to (3) when $p < 1$?

The answers to these questions are unknown, even for the forward light cone.

Proof of the Theorem Consider first the case when Γ is polygonal, i.e., Γ is generated by finitely many lines through 0. We may even suppose that the lines are linearly independent. For any other polygonal cone is a union of such cones. There exists a non-singular linear transformation which takes our cone's generators to the n unit coordinate vectors. The theorem is known on tubes over such cones. Now, if Γ_0 is a proper subcone of Γ, then one can find a polygonal cone between the two. Hence (1) and (2) follow.

In the case of (3), F is a Poisson integral of an L^p function on the full cone. So the conclusion (3) follows from standard results about Poisson integrals. For details, see [STW1], p. 119.

Chapter 2
More on Hardy Spaces

2.1 Hardy Spaces and Maximal Functions

Before proceeding we emphasize a few fundamental points. We have

Theorem 2.1.1 *Suppose that $F \in H^p(T_\Gamma)$, $p > 0$, where Γ is the first octant in $\mathbb{R}^n$. Then*

(1) *F has a nontangential limit almost everywhere in each variable. In particular,* $\lim_{\substack{y \in \Gamma \\ y \to 0}} F(x + iy) = F(x)$, *for almost every $x \in \mathbb{R}^n$. This is essentially unrestricted convergence (no subcone in y).*
(2) *$\int_{\mathbb{R}^n} |F(x + iy) - F(x)|^p \, dx \to 0$ as $y \in \Gamma$ tends to 0.*

Proof See Stein and Weiss, page 115. □

Remark 2.1.2 This result may fail for tubes over general cones. Unrestricted convergence of Poisson integrals may even fail for functions in $L^p(\mathbb{R}^n)$, $p > 1$. For example, if $\Gamma = C_n$ (the forward light cone in $\mathbb{R}^n$), $n \geq 3$, then there exists an $f \in L^p$, $p < \infty$, such that, for almost every x,

$$\limsup u(x, y) = \limsup \int_{\mathbb{R}^n} P(x - t, y) f(t) \, dt = \infty,$$

as $y \in \Gamma$ tends to 0 unrestrictedly.

However, if we use restricted convergence and if we use as a base for the cone one of the so-called domains of positivity described in the Vinberg–Koecher–Rothaus theorem, then for almost every $x \in \mathbb{R}^n$, $u(x, y) \to f(x)$ restrictedly. For details, see [STW2].

We of course used the result of Theorem 2.1.1 to prove the last theorem in the last chapter.

© The Author(s), under exclusive license to Springer Nature Switzerland AG 2023
S. G. Krantz, *The E. M. Stein Lectures on Hardy Spaces*, Lecture Notes in
Mathematics 2326, https://doi.org/10.1007/978-3-031-21952-8_2

In what follows, we concentrate on circular cones, although the results go over to domains of positivity.

Theorem 2.1.3 *Let* $\Gamma = C_n = \{y \in \mathbb{R}^n : y_1^2 - \sum_{j=2}^n y_j^2 > 0\}$.

(1) *Let* $\Gamma_0 \subseteq \Gamma$ *be a proper subcone,* $f \in L^p(\mathbb{R}^n)$, $1 \leq p \leq \infty$. *Let* $\mathcal{P}_y$ *be the Poisson kernel for* Γ. *Set*

$$u(x, y) = \mathcal{P}_y * f$$

and

$$Mf(x) \equiv \sup_{\substack{|x - \bar{x}| < cy \\ y \in \Gamma_0}} |u(\bar{x}, y)|.$$

Then

(a) $Mf < \infty$ *almost everywhere.*

(b) M *is weak type* $(1, 1)$.

(c) $\|Mf\|_{L^p} \leq A_p \|f\|_{L^p}$, $1 < p \leq \infty$.

Also, restricted limits exist almost everywhere.

(2) (a) *For any* $p < \infty$, *there exists an* $f \in L^p$ *such that* $\sup_{y \in \Gamma} |u(x, y)| = \infty$ *almost everywhere in* x.

(b) *There exists* $f \in L^\infty$ *with bounded support such that* $\lim_{\substack{y \to 0 \\ y \in \Gamma}} u(x, y)$ *does not exist almost everywhere.*

These results are due to Mary and Guido Weiss, Norman Weiss, and E. M. Stein. The proofs are rather long and involved. We first treat the positive results.

FIRST IDEA: If R is any rectangle centered at 0, form the maximal function:

$$f_R^*(x) = \sup_{\epsilon > 0} \frac{1}{m(\epsilon R)} \int_{\epsilon R} |f(x - t)| \, dt.$$

The estimates that we get for the operator $f \mapsto f_R^*$ are independent of the size and shape of R. In fact

$$\|f_R^*\|_{L^\infty} \leq \|f\|_{L^\infty}$$

and

$$m\{x : f_R^*(x) > \alpha\} \leq \frac{3^n}{\alpha} \|f\|_{L^1}.$$

These estimates hold uniformly when R is any compact, convex set.

SECOND IDEA: (This is an improvement of a previous idea about summing weak type 1 functions.)

Lemma 2.1.4 *Suppose that $f_j(x) \geq 0$, $j = 1, 2, \ldots$, satisfy $m\{x : f_j(x) > \alpha\} \leq 1/\alpha$, all $\alpha > 0$, all j. Also suppose $1 > c_j > 0$, $\sum c_j \log(1/c_j) = K < \infty$. We claim that $f = \sum c_j f_j$ is weak type 1 and*

$$m\{x : |f(x)| > \alpha\} \leq \frac{2K + 4}{\alpha}.$$

Proof Start by supposing that $K = 1$. Fix $\alpha > 0$. Let $f_j = f_j^1 + f_j^2 + f_j^3$ for each j, where

$$f_j^1 = \begin{cases} f_j & \text{when } f_j \leq \alpha/2 \\ 0 & \text{otherwise} \end{cases}$$

$$f_j^2 = \begin{cases} f_j & \text{when } \alpha/2 < f_j \leq \alpha/(2c_j) \\ 0 & \text{otherwise} \end{cases}$$

$$f_j^3 = \begin{cases} f_j & \text{when } f_j > \alpha/(2c_j) \\ 0 & \text{otherwise} \end{cases}$$

Write $f = f^1 + f^2 + f^3$ accordingly.

Notice that

$$m\{x : f^3 > 0\} \leq \sum m \left\{ x : f_j(x) > \frac{\alpha}{2c_j} \right\}$$

$$\leq \sum \frac{2c_j}{\alpha}$$

$$\leq \frac{2}{\alpha}.$$

So

$$m\{x : f(x) > \alpha\} \leq \frac{2}{\alpha} + m\{x : f^1 + f^2 > \alpha\}$$

$$\leq m \left\{ x : f^1 > \frac{\alpha}{2} \right\} + m \left\{ x : f^2 > \frac{\alpha}{2} \right\} + \frac{2}{\alpha}.$$

But of course $|f^1| \leq \sum c_j(\alpha/2) \leq \alpha/2$. Hence

$$m \left\{ x : f^1 > \frac{\alpha}{2} \right\} = m(\emptyset) = 0.$$

Finally,

$$m\left\{x : f^2 > \frac{\alpha}{2}\right\} \le \frac{2}{\alpha} \int f^2 \, dx$$

by Chebyshev's inequality. And this is

$$\le \frac{2}{\alpha} \sum c_j \int_{\alpha/2 < f_j \le \alpha/(2c_j)} f_j \, dx$$

$$\le \sum \frac{2c_j}{\alpha} \int_{\alpha/2}^{\alpha/(2c_j)} \frac{1}{\lambda} \, d\lambda$$

$$= \sum \frac{2}{\alpha} c_j \log(1/c_j).$$

$\square$

Now let $\bar{y} \in \Gamma_0$, $\bar{y} = (\bar{y}_1, 0, \ldots, 0)$. We claim that $\mathcal{P}_{\bar{y}}(x - t) \le C\mathcal{P}_{\bar{y}}(x)$ if $|t| \le c|\bar{y}|$. Recall that

$$\mathcal{P}_y(x) = \frac{c(y, y)^{n/2}}{[((x, x) - (y, y))^2 + 4(x, y)^2]^{n/2}},$$

where $(y, y) \equiv y_1^2 - y_2^2 - \cdots - y_n^2$. Thus, if we restrict attention to y of the form $y = (y_1, 0, \ldots, 0)$, then, e.g.,

$$\mathcal{P}_{(1,0,\ldots,0)}(x) = \frac{C}{[((x, x) - 1)^2 + 4x_1^2]^{n/2}}. \tag{$*$}$$

If $(*)$ were rapidly decreasing in all directions in the x variable at the same rate, then the proof of the theorem would be standard. One can readily see that the x_1 variable is distinguished from the others. One has that $\mathcal{P}_{(1,0,\ldots,0)}$ vanishes slowly in the x_1 direction, but in some sense this is a relatively small part of the space. In other words, if one sketches Γ, one can see that if $x \to \infty$ along the axis (the x_1 direction) of the cone, then $\mathcal{P}_{(1,0,\ldots,0)} \sim |x|^{-n}$, but as $x \to \infty$ in the horizontal directions, $\mathcal{P}_{(1,0,\ldots,0)} \sim |x|^{-2n}$. See Fig. 2.1. We remark that considerations such as these have motivated the study of non-isotropic spaces.

We break up the x-space in a manner which respects the above observations: Let $x = (x_1, x_2, \ldots, x_n)$ and $r = |\sum_{j=2}^{n} x_j^2|^{1/2}$. Then $|x_1| + r$ is essentially $|x| =$ distance from 0. Also $|x_1| - r$ is essentially the distance from x to the edge of the cone.

We write $\mathbb{R}^n = \cup S_{i,j}$, where

$$S_{i,j} = \begin{cases} 2^{i-1} \le |x_1| + r \le 2^i & \text{if } i = 1, 2, \ldots \\ 2^{j-1} \le ||x_1| - r| \le 2^j & \text{if } j = 1, 2, \ldots \end{cases}$$

Fig. 2.1 The x-space $\mathbb{R}^n$ **x**

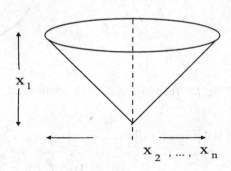

$$S_{0,j} = \begin{cases} & |x_1| + r \le 1 \\ 2^{j-1} \le ||x_1| - r| \le 2^j & \text{if} \quad j = 1, 2, \ldots \end{cases}$$

$$S_{i,0} = \begin{cases} 2^{i-1} \le |x_1| + r \le 2^i & \text{if} \quad i = 1, 2, \ldots \\ ||x_1| - r| \le 1 \end{cases}$$

$$S_{0,0} = \{|x_1| + r \le 1$$

Clearly $\mathbb{R}^n = \cup S_{i,j}$. We claim that

$$\mathcal{P}_{(1,0,\ldots,0)}(x) \sim 2^{-n(i+j)} \text{ on } S_{i,j}.$$

Now $|(x, x)| = |(x_1 + r) \cdot (x_1 - r)| \sim 2^{i+j}$ on S_{ij}. Also, we may as well assume that $j \le i + 1$ otherwise $S_{i,j}$ is empty. Therefore

$$\mathcal{P}_{(1,0,\ldots,0)}(x) \sim \frac{C}{[(2^{i+j} - 1)^2 + 4x_1^2]^{n/2}}$$

$$\le \frac{C}{(2^{i+j})^{2 \cdot n/2}}$$

$$= \frac{C}{2^{n(i+j)}}.$$

Note that, if $|x_1| > r$, then

$$|x_1| \sim |x_1| + r \sim 2^i$$

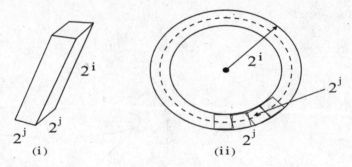

Fig. 2.2 The dotted "circle" is $(n-2)$-dimensional

and if $|x_1| \le r$, then

$$|x_1| \sim |x_1| + r - (r - |x_1|) \sim 2^i - 2^j \sim 2^i .$$

Hence

$$\mathcal{P}_{(1,0,\ldots,0)}(x) \ge \frac{C}{[(2^{i+j})^2 + 4 \cdot 2^{2i}]^{n/2}} \sim \frac{C}{2^{(i+j)n}} .$$

This proves the claim.
Thus

$$u(x, (1,0,\ldots,0)) = \mathcal{P}_{(1,0,\ldots,0)} * f \le C \sum 2^{-n(i+j)} \int_{S_{ij}} f(x-t)\,dt .$$

Clearly $m(S_{i,j}) \sim 2^{i(n-1)+j}$. So

$$u(x, (1,0,\ldots,0)) \le C \sum 2^{-(i+(n-1)j)} \frac{1}{m(S_{ij})} \int_{S_{ij}} f(x-t)\,dt . \qquad (\star)$$

Now unfortunately each S_{ij} is shaped like a sleeve and there is no maximal theorem for sets of that shape. Hence we must write each S_{ij} as a union of rectangles. Each rectangle will have dimensions $2^i \times 2^j \times \cdots \times 2^j$. If we write $i - j = k$, then it is easy to see that this will require essentially $C_1 \cdot 2^{k(n-2)}$ rectangles. See Fig. 2.2.

Now let us consider a schematic of S_{ij}. It is a sleeve-shaped section of $\mathbb{R}^n$, such that the sleeve is modeled on a cone. Consider Fig. 2.3

Thus

$$\text{no. of rectangles} \times \text{volume of rectangle}$$

$$\approx 2^{(i-j)(n-2)} \times 2^{i+j(n-1)}$$

$$\approx 2^{i(n-1)+j}$$

$$\approx m(S_{ij}) .$$

Fig. 2.3 The sleeve-shaped
section of $\mathbb{R}^n$

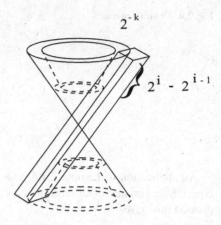

Now let ℓ, j be fixed, $0 \leq \ell \leq C_1 2^{k(n-2)}$, and let f_ℓ^* be the maximal function corresponding to the ℓth rectangle of S_{ij}. Clearly $f_{ij}^* \leq \sum_{\ell=1}^{N} f_\ell^*$, where $N \approx C_1 2^{k(n-2)}$ and

$$f_{ij}^* = \sup_\epsilon \frac{1}{m(\epsilon S_{ij})} \int_{\epsilon S_{ij}} f(x+t)\, dt .$$

We will use the following version of Lemma 2.1.4:

Lemma 2.1.5 *Suppose the f_j are as in Lemma 2.1.4. Then*

$$m\left\{ x : \frac{1}{N} \sum_{j=1}^{N} f_j(x) > \alpha \right\} \leq \frac{2\log(N+2)}{\alpha} \ , \ \alpha > 0.$$

Proof Trivial from Lemma 2.1.4.   □

Thus

$$m\{ x : f_{ij}^*(x) > \alpha \} \leq \frac{C\log(N+2)}{\alpha}$$

and, by $(\star)$, a previous observation, and the lemma, this is

$$\leq \frac{C\log(C_1 2^{k(n-2)})}{\alpha}$$

$$\leq \frac{C + Ck(n-2)}{\alpha} .$$

Fig. 2.4 Conical approach

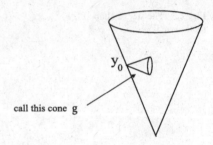

call this cone g

An application of Lemma 2.1.4 now shows that $\sum 2^{-(i+(n-1)j)} f_{ij}^*(x)$ is weak type 1. Standard interpolation arguments now give L^p estimates for $p > 1$. A simple dilation now gives

$$\sup_{\substack{y \in \mathbb{R} \\ y>0}} u(x, (y, 0, \ldots 0)) \le C \sum 2^{-(i+(n-1)j)} f_{ij}^*(x)$$

and the function on the right is weak type 1 or L^p according as $f \in L^1$ or $f \in L^p$, $p > 1$. By standard arguments, this estimate gives all the positive results of the theorem when the subcone Γ_0 is chosen to be the axis of Γ. For general Γ_0 the argument is the same, except that one must estimate $\mathcal{P}$ more carefully.

Now we indicate how the negative results are proved. [For details and greater generality, see [STN].

Let $f : \mathbb{R}^n \to \mathbb{C}$, $u(x, y) = \mathcal{P}_y * f$. What happens when $y \to 0$, but not through a proper subcone?

Let $d\mu_y = \mathcal{P}_y(x)\, dx$. If $y_0 \in b\Gamma$, $y_0 \neq 0$, write $y_0 = s y_0'$, $|y_0'| = 1$, $s > 0$. If we approach y_0 through a conical neighborhood of ys as shown in Fig. 2.4, we claim that

$$\mu_y \to \text{the one-dimensional Poisson kernel}.$$

We now make more precise the sense in which this is true.

Suppose that L is a directed line through 0 with $e \in \mathbb{R}^n$ the unit vector determining L. Suppose that $d\mu$ is a finite measure on $\mathbb{R}^1$. We claim that we may concentrate $d\mu$ on L in such a way that this determines a singular measure on $\mathbb{R}^n$. Namely, suppose that $\varphi \in C_0(\mathbb{R}^n)$. Let

$$d\mu^L(\varphi) = \int \varphi \, d\mu^L \equiv \int_{-\infty}^{\infty} \varphi(es) \, d\mu(s).$$

A measure $d\nu$ is said to be *supported on L* if, whenever $\varphi \in C_0(\mathbb{R}^n)$, $(\operatorname{supp} \varphi) \cap L \neq \emptyset$, then $\int \varphi \, d\nu = 0$.

Proposition 2.1.6 *We have that $d\nu$ is supported on L if and only if $d\nu = (d\mu)^L$, some $d\mu$ on $\mathbb{R}^1$.*

Proof Let dv be a measure on $\mathbb{R}^n$. Define

$$\widehat{v}(\xi) = \int_{\mathbb{R}^n} e^{2\pi i x \cdot \xi} \, dv(x).$$

Then dv is supported on L if and only if $\widehat{v}(\xi_1) = \widehat{v}(\xi_2)$ whenever $\xi_1 - \xi_2 \perp L$. But then $\widehat{v}$ restricted to any line parallel to L is the same so $\widehat{v}$ can be thought of (via projection) as a function on the line. It follows that if we define $d\mu$ on $\mathbb{R}^1$ by $(d\mu)\widehat{} = \widehat{v}$ (in the above sense), then $dv = (d\mu)^L$.

With this definition of $(d\mu)^L$, we let $dv_y = \mathcal{P}_y(t) \, dt$ on $\mathbb{R}^1$ and $(d\eta_y)^L$ be the concentration of $d\eta_y$ on a line L through $0 \in \mathbb{R}^n$ (the x-space). Let μ_y be as in the discussion preceding the proposition. We wish to show that $d\mu_y \to (d\eta_S)^L$ as $y \to y_0$, where the choice of L depends on the initial choice of y_0. This will clearly show that unrestricted convergence cannot work, since an $f \in L^p(\mathbb{R}^n)$ looks completely different in different directions. □

What the above argument purports to show, in an admittedly vague way, is that the problem of unrestricted convergence for Poisson integrals is equivalent to the problem of differentiation of the integral over long and skinny rectangles. Therefore the following considerations will complete our counterexample.

2.2 More Maximal Functions

We define the hyper-maximal function M_H by

$$M_H f(x) = \sup_{R \in \mathcal{R}} \frac{1}{m(R)} \int_R f(x, t) \, dt,$$

where

$$\mathcal{R} = \{\text{all rectangles centered at } 0\}.$$

Let S be a set. We say that a point $x \in S \subseteq \mathbb{R}^n$ is a point of *hyperdensity* if

$$\lim_{\substack{\text{diam } \rho \to 0 \\ \rho \in R}} \frac{m(S \cap \rho)}{m(\rho)} = 1.$$

Theorem 2.2.1 *We have:*

(1) *For all $p < \infty$, there exists an $f \in L^p(\mathbb{R}^2)$ such that $M_H f = \infty$ almost everywhere.*

(2) *There exists a set E, $m(E) > 0$, such that no point of E is a point of hyperdensity.*

Proof (1) For $p < 2$, simply let

$$f(x) = \begin{cases} |x|^{-1} & \text{if} \quad |x| < 1 \\ 0 & \text{if} \quad |x| \geq 1. \end{cases}$$

Then $f \in L^p$ for all $p < 2$. Let $x \in \mathbb{R}^n$. For rectangles ρ we choose long, skinny rectangles passing through x, 0 such that

$$\frac{1}{\rho} \int_\rho f(x+t)\, dt \sim \frac{1}{2N} \int_{-N}^{N} f(x+t)\, dt.$$

Since f restricted to a line through 0 is not in $L^1(\mathbb{R}^1)$, this gives **(1)** for $p < 2$. The proof for $p \geq 2$ we omit.

The proof of **(2)** follows from a construction of Nikodym in *Fundamenta Mathematicae* 10(1927), 116–168. The observation that **(2)** follows from Nikodym's work is contained in a remark of Zygmund at the end of that paper. Nikodym showed that

There exists a set $E \subseteq$ unit square $\subseteq \mathbb{R}^2$, $m(E) = 1$, such that: At any point in the set there exists a ray (half-infinite line) such that the ray does not intersect E.

Now pick $E_0 \subseteq E$ closed, $m(E_0) > 0$. One can easily see that

$$\liminf_{\text{diam}\,\rho \to 0} \frac{1}{m(\rho)} \int_\rho f(x+t)\, dt < \frac{1}{2} \quad \text{everywhere in } E_0,$$

but the ordinary density is 1 almost everywhere in E_0. So the limit of $\frac{1}{m(\rho)} \int_\rho f(x+t)\, dt$ does not exist almost everywhere as diam $\rho \to 0$, $\rho \in \mathcal{R}$.

By taking the product of the Nikodym–Zygmund example with an appropriate ball, one can obtain a counterexample in any $\mathbb{R}^n$. Problems of differentiation are discussed in Busemann–Feller, *Fundamenta Mathematica*, 1934 using the Kakeya needle problem. □

This completes our discussion of the negative result in Theorem 2.1.3, modulo the proof that $d\mu_y \to (d\eta_S)^L$. We now do this.

Proposition 2.2.2 *Let $\mathcal{P}_y$ be the Poisson kernel for T_Γ, Γ an arbitrary cone. Then*

$$\check{\mathcal{P}}_y(\xi) = e^{-2\pi |y \cdot \xi|}, \quad \xi \in \Gamma_* \cup -\Gamma_*.$$

Proof Recall that, by definition,

$$\mathcal{P}_y(x) = \frac{|C(x+iy)|^2}{C(2iy)}, \quad C(x+iy) = \int_{\Gamma_*} e^{2\pi i z \cdot t}\, dt, \quad z = x + iy.$$

So $C(x + iy)$, as a function of x, equals $(e^{-2\pi y \cdot t} \chi_{\Gamma^*}(t))\widehat{}$. Also $\overline{C(x + iy)} = (e^{2\pi y \cdot t} \chi_{-\Gamma^*}(t))\widehat{}$. As a result,

$$\left(|C(x + iy)|^2\right)^{\vee} = (e^{-2\pi y \cdot t} \chi_{\Gamma^*}(t)) * (e^{2\pi y \cdot t} \chi_{-\Gamma^*}(t)),$$

where of course $\vee$ is the inverse of $\widehat{}$.

As a result,

$$(\mathcal{P}_y(\cdot))^{\vee}(t) = \frac{(e^{-2\pi y \cdot t} \chi_{\Gamma^*}(t)) * (e^{2\pi y \cdot t} \chi_{-\Gamma^*}(t))}{C(2iy)}$$

$$= \frac{\int e^{-2\pi y(t-s)} \chi_{\Gamma^*}(t-s) e^{2\pi y \cdot s} \chi_{+\Gamma^*}(-s)\, ds}{\int_{\Gamma^*} e^{-4\pi y \cdot s}\, ds}$$

$$= \frac{\int e^{-2\pi y(s+t)} \chi_{\Gamma^*}(s+t) e^{-2\pi y \cdot s} \chi_{\Gamma^*}(s)\, ds}{\int_{\Gamma^*} e^{-4\pi y \cdot s}\, ds}.$$

Now if $t \in \Gamma^*$, then

$$\chi_{\Gamma^*}(s) = 1 \Rightarrow \chi_{\Gamma^*}(s) \cdot \chi_{\Gamma^*}(s+1) = 1.$$

So the last integral equals

$$e^{-2\pi y \cdot t} \frac{\int e^{-4\pi y \cdot s} \chi_{\Gamma^*}(s)\, ds}{\int e^{-4\pi y \cdot s} \chi_{\Gamma^*}(s)\, ds} = e^{-2\pi y \cdot t} = e^{-2\pi |y \cdot t|} \quad \text{since } y \in \Gamma, \ t \in \Gamma^*.$$

The proof for $t \in -\Gamma^*$ is similar. □

Remark 2.2.3 Such a nice representation for $\mathcal{P}_y$ may not hold in general.

In order to prove our result about convergence of measures on the forward light cone, it will be convenient to have another realization of it in $\mathbb{R}^3$.

Consider

$$\left\{ \begin{pmatrix} u_1 & u_3 \\ u_3 & u_2 \end{pmatrix} : u_1 > 0, u_1 u_2 - u_3^2 = \det(\ ) > 0 \right\}.$$

The change of variables

$$u_1 = y_1 + y_2$$
$$u_2 = y_1 - y_2$$
$$u_3 = y_3$$

takes this representation to the old one.

We remark that this is the Jordan algebra realization and notation of C_3. It happens to be easier to study linear transformations that preserve the cone in this new coordinate system.

Proposition 2.2.4 *Let Γ be any cone in $\mathbb{R}^n$, $T : \Gamma \to \Gamma$, T linear. Let $\mathcal{P}_y(x)$ be the Poisson kernel for Γ. Then*

$$\mathcal{P}_{Ty}(Tx) = (\mathcal{P}_y(x)) \cdot |\det T|^{-1} .$$

Proof It suffices to verify such a formula for the Cauchy kernel. That is, we show that

$$C(Tx + iTy) = C(Tz) = |\det T|^{-1} C(z) .$$

Recall that $C(z) = \int_{\Gamma^*} e^{2\pi i z \cdot t} \, dt$, so

$$C(Tz) = \int_{\Gamma^*} e^{2\pi i (Tz) \cdot t} \, dt = \int_{\Gamma^*} e^{2\pi i z \cdot (T^* t)} \, dt .$$

Now do the change of variable $s = T^* t$, $ds = |\det T^*| \, dt$. So

$$C(Tz) = |\det T|^{-1} \int_{\Gamma^*} e^{2\pi i z \cdot t} \, dt = |\det T|^{-1} C(z) .$$

That completes the proof.   □

It can be shown that the transformations which preserve C_3 are given by

$$\begin{pmatrix} x_1 & x_3 \\ x_3 & x_2 \end{pmatrix} \longmapsto a \begin{pmatrix} x_1 & x_3 \\ x_3 & x_2 \end{pmatrix} {}^t a,$$

where a is a 2×2 matrix. For instance, if a is diagonal, then the transformation becomes

$$T_\delta : \begin{pmatrix} x_1 & x_3 \\ x_3 & x_2 \end{pmatrix} \longmapsto \begin{pmatrix} \delta_1^2 x_1 & \delta_1 \delta_2 x_3 \\ \delta_1 \delta_2 x_3 & \delta_2^2 x_2 \end{pmatrix} \quad \text{if } a = \begin{pmatrix} \delta_1 & 0 \\ 0 & \delta_2 \end{pmatrix} .$$

Suppose now that

$$y = \begin{pmatrix} 1 & 0 \\ 0 & 1 \end{pmatrix} \in C_3 .$$

Then

$$\lim_{\delta_2 \to 0} Ty = \lim_{\delta_2 \to 0} \begin{pmatrix} \delta_1^2 & 0 \\ 0 & \delta_2^2 \end{pmatrix} = \begin{pmatrix} \delta_1^2 & 0 \\ 0 & 0 \end{pmatrix} \in \text{boundary of } C_3 .$$

Now we have a sufficient amount of machinery to check that $d\mu_y \mapsto d(\eta_S)^L$ with the notation that we introduced in the last section. Let φ be a test function, $\delta_1 = 1$. We will let $\delta_2 \to 0$. Then, recalling that $\mathcal{P}_{Ty}(x) = |\det T|^{-1}\mathcal{P}_y(T^{-1}x)$, we have (using the notation I for the matrix $\begin{pmatrix} 1 & 0 \\ 0 & 1 \end{pmatrix}$)

$$\int_{\mathbb{R}^3} \mathcal{P}_{Ty}(x)\varphi(x)\,dx = \int_{\mathbb{R}^3} \mathcal{P}_{T(I)}(T^{-1}x)\varphi(x)\,dx$$

$$= |\det T|^{-1} \int_{\mathbb{R}^3} \mathcal{P}_I(T^{-1}x)\varphi(x)\,dx$$

$$= \int \mathcal{P}_I(x)\varphi \begin{pmatrix} x_1 & \delta_2 x_3 \\ \delta_2 x_3 & \delta_2^2 x_2 \end{pmatrix} dx$$

by the obvious change of variable.

As $\delta_2 \to 0$, certainly

$$\varphi \begin{pmatrix} x_1 & \delta_2 x_3 \\ \delta_2 x_3 & \delta_2^2 x_2 \end{pmatrix} \to \varphi \begin{pmatrix} x_1 & 0 \\ 0 & 0 \end{pmatrix}.$$

Thus if φ is supported off the line through I, then the right side vanishes as $\delta_2 \to 0$ hence so does the left. But this says that $\lim_{\delta_2 \to 0} \mathcal{P}_{Ty}(\cdot)$ is supported on the line through I, that is

$$\mathcal{P}_{\tilde{y}}(\cdot) \to \text{(a measure supported on the line through } I)$$

$$\text{as } \tilde{y} \to \begin{pmatrix} 1 & 0 \\ 0 & 0 \end{pmatrix} \in \text{boundary } C_3.$$

Clearly that measure is $d(\eta_1)^L$ as desired.   □

Now we consider some speculations and open problems. We begin by recalling a few notions.

For Poisson integrals of L^p functions, restricted convergence appears to "work," while unrestricted convergence does not.

Let $C(z) = C(x + iy)$ be the Cauchy kernel. Then write

$$F(z) = \int_{\mathbb{R}^3} C(z - t)f(t)\,dt \ , \ f \in L^p(\mathbb{R}^2) \ , z = x + iy \ , y \in \Gamma.$$

We wish to consider the mapping

$$f \longmapsto \text{boundary values of } F.$$

[This is where the Hilbert transform arises classically.]

Fig. 2.5 Halfspace
determined by unit vector

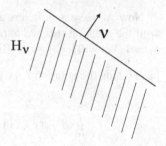

Define three operators arising from Γ^*:

- P_+: Given by multiplication by the characteristic function of the forward cone.
- P_-: Given by multiplication by the characteristic function of the negative cone.
- P_0: Given by multiplication by the characteristic function of the complement of the forward and negative cones.

More precisely, let

$$\widehat{P_+(f)} = \chi_{\Gamma^*}(t)\,\widehat{f}(t),$$

$$\widehat{P_-(f)} = \chi_{-\Gamma^*}(t)\,\widehat{f}(t),$$

$$\widehat{P_0(f)} = \widehat{f} - \widehat{P_+ f} - \widehat{P_- f}.$$

Question 1 Are P_+, P_-, P_0 bounded operators on L^p, $1 < p < \infty$?

Answer No, except when $p = 2$.

Question 2 Let $\mathcal{D}$ be the unit ball in $\mathbb{R}^2$, and define $(T_{\mathcal{D}} f)^{\widehat{\,}} = \chi_{\mathcal{D}} \cdot \widehat{f}$. Is $T_{\mathcal{D}}$ bounded on L^p?

Answer No, except when $p = 2$. (**Reference**: Charles Fefferman [FEF1].)

Fefferman's result also yields a "no" answer to Question 2 for $n > 2$ as we see from:

Theorem 2.2.5 (de Leeuw) *If χ_S is a multiplier on $\mathbb{R}^n$, then $S \cap (hyperplane)$ is a multiplier on $\mathbb{R}^{n-1}$.*

There is a parallel between the counterexample for the disc and that for unrestricted convergence.

Question 3 In 2 dimensions, consider a variety of half spaces at arbitrary angles. Each halfspace is uniquely determined by a unit vector. See the Fig. 2.5.

For each halfspace H_v we may construct a Hilbert transform by

$$(H_v f)^{\widehat{\,}} = i\chi_{H_v}\widehat{f} - i\chi_{-H_v}\widehat{f}.$$

One now has the following analogue to the Littlewood–Paley theory.

Let $f_1, f_2, \ldots$ be functions on $\mathbb{R}^2$. One can ask whether

$$\|(\sum_k |H_{v_k}(f_k)|^2)^{1/2}\|_{L^p} \leq C_p \|(\sum_k (f_k)^2)^{1/2}\|_{L^p}.$$

For $p = 2$ it is clear.

One can show (Yves Meyer) that, for those p for which Question 2 has a positive answer, then so does Question 3. Fefferman used the Kakeya needle construction to give a counterexample to Question 3.

PROBLEM 1 Does there exist $F \in H^p(T_\Gamma)$, $1 \leq p \leq \infty$, such that unrestricted limits fail to exist almost everywhere?

The problem has an "easy solution" when $p = 2$. Start with

$$f(x) = \begin{cases} |x|^{-1}/(\log[1/|x|])^{3/4} & \text{for } |x| \text{ small} \\ 0 & \text{for } |x| \text{ large}. \end{cases}$$

Clearly $f \in L^2(\mathbb{R}^2)$ and $C(f)(z)$ exists, $Cf \in H^2(T_\Gamma)$. There will be more details on this example later.

PROBLEM 2 What is the (possibly many) maximal function which controls the operators $\mathcal{P}_0$, $\mathcal{P}_-$, $\mathcal{P}_+$? More precisely, there presumably exists an operator $f \mapsto \mathcal{M}f$ such that

$$\mathcal{M}f \in L^p \Rightarrow \mathcal{P}_0 f \in L^p, \quad \text{etc.}$$

The classical maximal function which controls restricted convergence is too good because it is bounded for all L^p, $p > 1$. By PROBLEEM 1, the maximal function we are looking for cannot be bounded on L^2. We can say more, however, than merely that things seem to work for restricted convergence and do not seem to work for unrestricted convergence.

PROBLEM 3 Let $\Gamma = C_3$, $f \in L^p$, $u(x, y) = PI(f)$. Let Π be the hyperplane indicated in Fig. 2.6. Let $M_\Pi f(x) \equiv \sup_{y \in \Pi} u(x, y)$. Then

(1) $\|M_\Pi(f)\|_{L^p} \leq A_p \|f\|_{L^p}$, $1 < p \leq \infty$.
(2) $M_\Pi f < \infty$ almost everywhere if $f \in L \log L$.

Let $\mathcal{P}(x)$ denote $\mathcal{P}_{(1,0,\ldots,0)}(x)$. When we discuss restricted convergence, we consider $\{\mathcal{P}(x/\delta)\delta^{-3}\}_{\delta > 0}$. There is, as we discovered, a natural 2-parameter family of dilations associated with the problem, namely

$$\begin{pmatrix} x_1 & x_3 \\ x_3 & x_2 \end{pmatrix} \longmapsto \begin{pmatrix} \delta_1^2 x_1 & \delta_1 \delta_2 x_3 \\ \delta_1 \delta_2 x_3 & \delta_2^2 x_2 \end{pmatrix}.$$

Can we relate these considerations with

Fig. 2.6 A hyperplane
intersecting the cone

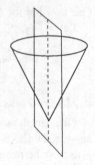

$$\mathcal{P}_y(x) = \mathcal{P}^1_{y_1}(x_1)\mathcal{P}^1_{y_2}(x_2) \cdots \mathcal{P}^1_{y_n}(x_n)\,?$$

PROBLEM 4 Let $\alpha_\delta : (x_1, x_2, x_3) \mapsto (\delta_1 x_1, \delta_2 x_2, \delta_1 \delta_2 x_3)$. Start with the unit cube $Q \subseteq \mathbb{R}^3$. Let

$$N(f) = \sup_{\substack{\delta_1 > 0 \\ \delta_2 > 0}} \frac{1}{m(\alpha_\delta(Q))} \int_{\alpha_\delta(Q)} |f(x - t)|\, dt\,.$$

It is easy to see that this is bounded on L^p, $p > 1$, by the strong differentiation theorem. **Question:** Is $Nf < \infty$ almost everywhere for $f \in L \log l$?

PROBLEM 5 (ZYGMUND) Consider a 2-parameter family of rectangles in 3-space, $\delta = (\delta_1, \delta_2)$; $R_\delta = R_{\delta_1, \delta_2}$ is a rectangle centered at 0, sides parallel to the axes, $R_{\delta_1, \delta_2} \subseteq R_{\delta'_1, \delta_2}$ whenever $\delta_1 < \delta'_1$, and $R_{\delta_1 \delta_2} \subseteq R_{\delta_1, \delta'_2}$ if $\delta_2 < \delta'_2$. Consider the maximal function $N(f)$ as above. Is $N(f) < \infty$ almost everywhere if $f \in L \log L$?

OPEN PROBLEM (Probably Quite Difficult) Let R be large. Take χ_R be the characteristic function of a disc of radius R. Let

$$(S_R f)\widehat{} \equiv \chi_R \cdot \widehat{f}\,.$$

Clearly, by Plancherel,

$$\|S_R f\|_{L^2} \le \|f\|_{L^2}$$

and $S_R f \to f$ in L^2. [This fails for L^p, $p \ne 2$, by a celebrated result of Charles Fefferman.] Does the convergence take place almost everywhere for $f \in L^2$? [Stein thinks not.][1] If $p < 2$, then, at least in the periodic case, it has been shown that the counterexample for the multiplier problem for the ball establishes that almost everywhere convergence fails for $p < 2$.

---

[1] This author believes this question to still be open.

2.3 Real Variable H^p

Now we return to H^p over the octant:

$$\Gamma = \{y = (y_1, y_2, \ldots, y_n) : y_j > 0 \text{ for all } j\}.$$

Assume for simplicity that $n = 2$. Then

$$H^p(T_\Gamma) = \{F \text{ holomorphic on } T_\Gamma : \sup_{y > 0} \int |F(x + iy)|^p \, dx \equiv \|F\|_{H^p}^p < \infty\}.$$

Define $F(x) = \lim_{\substack{y \to 0 \\ y \in \Gamma}} F(x + iy)$ in the appropriate sense. [We make no apologies for what follows as it has never been written down before and is admittedly vague and sketchy.]

Let η be an n-tuple of $+$ and $-$ signs, $\eta = (\eta_1, \eta_2, \ldots, \eta_n)$, each $\eta_j = \pm 1$. For each η, let

$$\Gamma_\eta = \{y : \eta_j y_j > 0, j = 1, \ldots, n\}.$$

Define the *real H^p space*

$$RH^p = \{F : F \text{ is a tempered distribution on } \mathbb{R}^n \text{ and } F = \sum_\eta F_\eta(x)$$

where each F_η is an H^p function on $T_{\Gamma_\eta}\}.$

[One can check that if $F \in H^p(T_\Gamma)$, then $\lim_{\substack{y \to 0 \\ y \in \Gamma}} F(x + iy)$ exists in the sense of tempered distributions.] Also, one can show that, for $p \geq 1$, $\widehat{F}$ is a function. We now propose a second definition of real H^p.

We will be discussing multiply harmonic functions u and, in particular, $u(x, y) = PI\, f$. We shall always assume that

$$|u(x, y)| \leq A y_1^{-a_1} \cdot y_2^{-a_2} \cdot \cdots \cdot y_n^{-a_n} \quad a_j > 0 \text{ for all } j. \tag{$*$}$$

This we do without loss of generality since $F \in H^p$, $p < \infty$ implies that

$$|F(x + iy)| \leq A(y_1 \cdot y_2 \cdot \cdots \cdot y_n)^{-1/p}.$$

We will let, for $f \in L^p(\mathbb{R}^n)$, $u^\eta f$ denote the Poisson integral of f over T_{Γ_η}. We will also write Γ for $\Gamma_{(1,1,\ldots,1)}$ and u for $u^{(1,1,\ldots,1)}$.

Definition 2.3.1 We say that $\{u^\eta\}$ is in real variable H^p if

(1) $u^\eta(x, y)$ are uniformly in L^p, $y \in \Gamma_\eta$.

(2) For each j,

$$u_{(x,y)}^{(\eta_1,\eta_2,...,\eta_{j-1},1,\eta_{j+1},...,\eta_n)} + i u_{(x,y)}^{(\eta_1,\eta_2,...,\eta_{j-1},-1,\eta_{j+1},...,\eta_n)}$$

is holomorphic in $z_j = x_j + iy_j$.

(3) There exist $a_j > 0$ such that

$$|u^\eta(x,y)| \le A \prod_{j=1}^n y_j^{-a_j} , \text{ all } \eta .$$

We wish to be able to pass back and forth between these two definitions of real H^p. Suppose that u is a given multiply harmonic function on T_Γ. Let

$$u^{(1,1)}(x,y) = u(x,y) , \quad \widehat{(u^{(1,1)})_y}(\xi) = \widehat{u}(\xi,y)$$

(we only take the Fourier transform in the first variable). Also

$$u^{(-1,1)}(x,y) = (\text{Hilbert transform})_{x_1} u(x,y) = H^1 u(x,y)$$

so that

$$\widehat{H^1 u}(\xi,y) = \frac{1}{i}\text{sgn } \xi_1 \widehat{u}(\xi,y) .$$

Furthermore,

$$u^{(1,-1)}(x,y) = H^2 u(x,y) ,$$

where

$$\widehat{H^2 u}(\xi,y) = \frac{1}{i}\text{sgn } \xi_2 \widehat{u}(\xi,y) .$$

Also,

$$u^{(-1,-1)}(x,y) = H^1 H^2 u(x,y) ,$$

where

$$\widehat{H^1 H^2 u}(x,y) = (-1)\text{sgn } \xi_1 \text{sgn } \xi_2 \widehat{u}(\xi,y) .$$

We claim that

$$\left[u^{(1,1)} - u^{(-1,-1)} \right] + i \left[u^{(-1,1)} + u^{(1,-1)} \right]$$

is holomorphic and supported in the first quadrant (i.e., its Fourier transform is so supported).

Consider, for example, the quadrant where $\xi_1 > 0$, $\xi_2 < 0$. Then the Fourier transform of this function on this quadrant is

$$(1-1) + i\left(\frac{1}{i}(1-1)\right)\widehat{u}(\xi, y) = 0.$$

The analyticity of this Fourier transform is easy.

We let

$$F = (I + iH^1) \cdot \cdots \cdot (I + iH^n)u.$$

This will be a sum of u^ns when expanded. Let

$$F_\eta(x + i\eta y) = (I + i\eta_1 H^1)(I + i\eta_2 H^2) \cdot \cdots \cdot (I + i\eta_n H^n)u.$$

Given a $\{F_\eta(z)\}$, we may define

$$u^{\eta'}(x, y) = \sum \sigma(\eta', \eta) F_\eta(x + i\eta y),$$

where

$$\sigma(\eta', \eta) = \prod_{\eta'_k = -1} (i^{\eta'_k} \cdot \eta_k).$$

Theorem 2.3.2 *We have:*

(a) *For $1 < p < \infty$, one has $f \in \mathbb{R}H^p$ if and only if $f \in L^p(\mathbb{R}^n)$.*

(b) *It holds that $f \in \mathbb{R}H^1$ if and only if $f \in L^1$ and all*
$$H^\eta \equiv H^{(\eta_1)} \cdot \cdots \cdot H^{(\eta_n)} f \text{ are in } L^1.$$

Observe that

$$H^{(\eta_k)} = \begin{cases} I & \text{on the } k^{\text{th}} \text{ variable if } \eta_k = 1 \\ H^k & \text{on the } k^{\text{th}} \text{ variable if } \eta_k = -1. \end{cases}$$

The above theorem is a consequence of the theory of tubes over octants so we omit it.

Remarks

(1) If $\{F_\eta\}$ with $F_\eta \in H^p(T_{\Gamma_\eta})$ and $\{G_\eta\}$ with $G_\eta \in H^p(T_{\Gamma_\eta})$ satisfy

$$\sum_\eta F_\eta(x) = \sum_\eta G_\eta(x),$$

where $F_\eta(x)$ and $G_\eta(x)$ are the boundary values of F_η, G_η in the distribution sense, then $F_\eta = G_\eta$.

(2) We may reformulate some of the above in a group context. For the two-element group $\{-1, 1\}$ under multiplication, the dual group is the same group. The pairing $\langle \eta', \eta \rangle$ is specified by

$$\langle \eta'_k, \eta_k \rangle = +1 \ \text{ if } \eta_k = \eta'_k = 1,$$

$$\langle \eta'_k, \eta_k \rangle = -1 \ \text{ if } \eta'_k = -1, \eta_k = 1,$$

$$\langle \eta'_k, \eta_k \rangle = +1 \ \text{ if } \eta_k = \eta'_k = -1,$$

$$\langle \eta'_k, \eta_k \rangle = -1 \ \text{ if } \eta'_k = 1, \eta_k = -1.$$

Now if we let H denote the ordinary Hilbert transform, then we define

$$\tilde{H}^{\eta_k} = I \ \text{ if } \eta_k = 1$$

and

$$\tilde{H}^{\eta_k} = iH \ \text{ if } \eta_k = -1.$$

If u is multiply harmonic in T_Γ, set

$$F = \sum_\eta \tilde{H}^\eta u.$$

Then one has

$$F_{\eta'}(x) = \sum_{\eta'} \langle \eta', \eta \rangle H^\eta(u)(x, 0)$$

and

$$H^\eta(u)(x) = \frac{1}{2^n} \sum \langle \eta, \eta' \rangle F_{\eta'}.$$

Here the Fourier transform of $F_{\eta'}$ is supported in Γ whereas the Fourier transform of $H^\eta u$ is supported in $\Gamma_{\eta'}$.

If u is multiply harmonic in T_Γ, we set

$$S(u) = \left(\int_{\Gamma(x_n) \times \cdots \times \Gamma(x_1)} \cdots \int |\nabla_n \cdots \nabla_2 \nabla_1 u(z')|^2 \, dz'_1 dz'_2 \cdots dz'_n \right)^{1/2}.$$

Theorem 2.3.3 *Assuming that $u \in RH^p$, we have*

$$\|S(u)\|_{L^p} \approx \|u\|_{RH^p}.$$

[**n.b.** We assume here that $u \in RH^p$ to eliminate consideration of functions which depend on fewer than n variables and functions with constants added on.]

Proof We make the following reduction: It suffices to show that $F \in H^p(T_\Gamma)$ if and only if $\|SF\|_{L^p} \approx \|F\|_{L^p}, 0 < p < \infty$.

Note that $S(u)$ is the same as $S(u^*)$, where u^* is any conjugate of u.

Next we suppose that $n = 2$.

Now suppose that $F \in H^p(T_\Gamma), 0 < p < \infty$. Then

$$\|F\|_{H^p} = \left(\int_{\mathbb{R}^2} |F(x_1, x_2)|^p \, dx_1 dx_2 \right)^{1/p}.$$

This is true because

$$\|F\|_{H^p} = \left(\sup_{\substack{y_1 > 0 \\ y_2 > 0}} \int |F(x_1 + iy_1, x_2 + iy_2)|^p \, dx_1 dx_2 \right)^{1/p}$$

$$\geq \left(\int_{\mathbb{R}^2} |F(x_1, x_2)|^p \, dx_1 dx_2 \right)^{1/p}.$$

The reverse inequality follows by harmonic majorization:

$$|F(x_1 + iy_1, x_2 + iy_2)|^p \leq \int P_{y_1 y_2}(t_1, t_2) |F(x_1 - t_1, x_2 - t_2)|^p \, dt_1 dt_2.$$

We make one more observation: If $u : \mathbb{R}^2 \times i\mathbb{R}^2 \to \mathcal{H}$, $\mathcal{H}$ a Hilbert space, and u is harmonic, then

$$\Delta(|u|^2) \approx 2|\nabla u|^2.$$

Now, for $0 < p < \infty$, let $F \in H^p(T_{\text{first quadrant}})$.

In what follows, we let subscripts denote the taking of the gradient in the indicated variable.

Then

$$S(F)(x_1, x_2) = \left(\iint_{\Gamma(x_1)\Gamma(x_2)} |F_{12}(z_1', z_2')|^2 \, dz_1' dz_2' \right)^{1/2}.$$

Now let

$$\mathcal{H} = \{f(z_2) : f \text{ is holomorphic in } \Gamma(x_2), \|f\| = \int_{\Gamma(x_2)} |f|^2 \, dz_2^{1/2} < \infty\}.$$

Let $z_1 \mapsto G(z_1)$ be a holomorphic function with values in $\mathcal{H}$, and

$$G(z_1) = G^{z_2}(z_1) \equiv F_2(z_1, z_2).$$

Then

$$\int \left(\int_{\Gamma(x_1)} \|G_1(z_1')\|^2 \, dz_1' \right)^{1/2} dx_1 \approx \int |G(x_1)|^p \, dx_1 \qquad (\star)$$

by the area theorem. But the left side of $(\star)$ is

$$= \int \left(\iint_{\Gamma(x_1)\Gamma(x_2)} |F_{12}(z_1', z_2')|^2 \right)^{1/2} dx_1$$

$$\approx \int_{\Gamma(x_2)} \left(\int |F_2(x_1, z_2')|^2 dz_2' \right)^{p/2} dx_1 .$$

Now integrate with respect to x_2 to get

$$\|S(F)\|_p^p = \int \left(\int_{\Gamma(x_2)} |F_2(x_1, z_2')|^2 \, dz_2' \right)^{p/2} dx_1 dx_2$$

$$= \int \left(\int \left(\int_{\Gamma(x_2)} |F_2(x_1, z_2')|^2 \, dz_2' \right)^{p/2} dx_2 \right) dx_1 .$$

If we think of the integral, for each fixed x_1, as a function $z_2 \mapsto F(x_1, z_2)$, then this

$$\approx \int \left(\int |F(x_1, x_2)|^p \, dx_2 \right) dx_1 ,$$

as desired.   □

Now we give some unsolved (and solved) problems.

Problem 1 Suppose that F is holomorphic on T_Γ. Prove that almost everywhere on the distinguished boundary the following are equivalent:

(a) The function F has an unrestricted limit at $(x_1, x_2, \ldots, x_n)$.
(b) The area integral is finite at $(x_1, x_2, \ldots, x_n)$.

Problem 2 Let u be multiply harmonic. Prove that almost everywhere the following are equivalent:

(a) The function u has an unrestricted limit at $(x_1, x_2, \ldots, x_n)$.
(b) We have $S(u) < \infty$ at $(x_1, x_2, \ldots, x_n)$.

A positive solution to Problem 2 would yield the following: Let u be multiply harmonic on T_Γ and suppose that, for $E \subseteq \mathbb{R}^n$, $|E| > 0$, we know that u has unrestricted limits almost everywhere in E. Then so do the $2^n - 1$ conjugates of u.

Remark 2.3.4 In $\mathbb{R}^2$, there exists a u which is multiply harmonic and has restricted limits almost everywhere, but none of its conjugates does. [This yields no contradiction regarding the above.]
If $f \in L^1$, then we know that, with $u = PIf$, restricted limits exist almost everywhere. However, there exists an f, $f \notin L \log L$, but $f \in L(\log L)^{1-\epsilon}$, such that with $v(x, y) = Q_{y_1} Q_{y_2} * f$ we have $\lim_{y_1 = y_2 \to 0} v(x, y)$ does *not* exist almost everywhere. This was stated in 1939 by Sokol-Sokolowsky but there is an error in his proof.
But in Stein, limits of sequences of operators, a proof is given. There is also an error in Stein's proof, for it only gives the result when $\epsilon > 1/2$.[2]

Remark 2.3.5 Assume that $f \in L \log L$. Then we get unrestricted convergence. More precisely, if we are in $\mathbb{R}^1$ (the periodic case) and $f \in L \log L$, then $F_\eta \in H^p$, $p < 1$ so unrestricted limits exist.

Note Cotlar claimed to have solved Problem 2 by real variable methods (the original proof was by complex methods). There was an error. But Fefferman claims to have a proof.

Problem 2′ [This is a global analogue of **Problem 2**.] Let u be multiply harmonic in T_Γ,

$$u^*(x) = \sup_{z_j' \in \Gamma_j(x_j)} |u(z')| .$$

Suppose that u is admissible at ∞. Then

$$\|S(u)\|_{L^p} \approx \|u^*\|_{L^p} , \quad 0 < p < \infty .$$

[This is closely related to the proof of Problem 2.]

Problem 3 Suppose that f is a tempered distribution on $\mathbb{R}^n$, $u = MPIf$. Say that $n = 2$. Let $0 < p < \infty$. Are any of the following equivalent?

(a) $u^* \in L^p$.

---

[2] The elegant paper [SAW] generalizes Stein's theorem and corrects some of the slips in Stein's original paper.

(b) For $\varphi^1, \varphi^2 \in \mathcal{S}(\mathbb{R}^1)$,

$$\sup_{\substack{\epsilon_1 > 0 \\ \epsilon_2 > 0}} |(\varphi^1_{\epsilon_1} \varphi^2_{\epsilon_2}) * f(x)| \in L^p,$$

where $\varphi^1 = \varphi^1(x_1)$, $\varphi^2 = \varphi^2(x_2)$.

(b') Same as **(b)**, but for just one choice of φ^1, φ^2.

(c) For $\varphi \in \mathcal{S}(\mathbb{R}^2)$ a function of 2 variables, let $\varphi_{\epsilon_1 \epsilon_2}(x_1, x_2) = \epsilon_1 \epsilon_2 \varphi(x_1/\epsilon_1, x_2/\epsilon_2)$. Then we have

$$\sup_{\substack{\epsilon_1 > 0 \\ \epsilon_2 > 0}} |\varphi_{\epsilon_1 \epsilon_2} * f(x)| \in L^p.$$

Problem 4 Assume that Problem 3 has an affirmative answer. Let

$$^*H^p = \{f, \text{ or all } u : \mathbf{a}, \mathbf{b}, \mathbf{b'}, \text{ or } \mathbf{c} \text{ holds}\}.$$

Is $^*H^p = RH^p$? **Remark:** Presumably $RH^p \subseteq {}^*H^p$ in any case.

Recall now the discussion following Problem 1 in Sect. 2.2. We had Γ a circular cone, $F \in H^p(T_\Gamma)$, $p = 2$. Unfortunately this example was *completely wrong*.

Let us consider now the problem of passing between the periodic and non-periodic situations. There exists a map

$$\mathbb{R}^n \longrightarrow \mathbb{R}^n / \Lambda \equiv T_n \text{ (the torus)},$$

where $\Lambda = \{(m_1, m_2, \ldots, m_n) : m_j \in \mathbb{Z}\}$. There are basically two ways of passing from non-periodic to periodic f:

(1) Given $f : \mathbb{R}^n \to \mathbb{C}$, let

$$f_p(x) = \sum_{m \in \Lambda} f(x + m).$$

(2) Write

$$\widehat{f}(\xi) = \int_{\mathbb{R}^n} e^{-2\pi i x \cdot \xi} f(x) \, dx$$

and

$$f(x) = \int_{\mathbb{R}^n} \widehat{f}(\xi) e^{2\pi i x \cdot \xi} \, d\xi.$$

Recall the Poisson summation formula:

$$\sum f(x+m) = \sum \widehat{f}(m)e^{2\pi i m \cdot x} .$$

Now suppose that $f \in L^1(\mathbb{R}^n)$. Define

$$f_p(x) = \sum_{m \in \Lambda} f(x+m) \sim \sum \widehat{f}(m)e^{2\pi i m \cdot x} ,$$

where the convergence is in L^1(fundamental cube). Here

$$(\text{fundamental cube}) = \left\{ x \in \mathbb{R}^n : |x_j| < \frac{1}{2} \right\} .$$

One has

$$\| f_p \|_{L^1} = \int_{\substack{(\text{fundamental} \\ \text{cube})}} \sum_{m \in \Lambda} |f(x+m)| \, dx = \sum_m \int_{\substack{(\text{fundamental} \\ \text{cube})}} |f(x)| \, dx = \| f \|_{L^1}$$

and

$$\widehat{f}(m) = \int_{\substack{(\text{fundamental} \\ \text{cube})}} f_p(x)e^{-2\pi i m \cdot x} \, dx .$$

Corollary 2.3.6 *Suppose that f is given, $|f| \le A(1+|x|)^{-n-\epsilon}$. Assume that*

$$|\widehat{f}(\xi)| \le A(1+|\xi|)^{-n-\epsilon} ,$$

so that both f and $\widehat{f}$ are continuous. Then

$$\sum_{m \in \Lambda} f(m) = \sum_{m \in \Lambda} \widehat{f}(m) .$$

2.4 Some Thoughts on Summability

As usual, we will write $f \sim \sum a_m e^{2\pi i m \cdot x}$, where $a_m = \int_{T^n} f(x)e^{-2\pi i m \cdot x} \, dx$. Here we assume that f is given on T^n or identifying f with f_p. One has that f is a tempered distribution on T^n if and only if $|a_m| \le A(1+|m|)^k$, some k, some A.

If $\varphi \sim \sum b_m e^{2\pi i m \cdot x}$, then

$$\int f\overline{\varphi} \, dx = \sum_{m \in \Lambda} a_m \overline{b}_m .$$

We wish to think about summability. In particular, we would like to study

$$\sum a_m e^{-2\pi|m|y} e^{2\pi im \cdot x} \ , \ y > 0 \, .$$

Let

$$\widetilde{P}_y(x) = \sum e^{-2\pi|m|y} e^{2\pi im \cdot x} \, .$$

FACT We have that

$$\widetilde{P}_y(x) = \sum_m P_y(x + m) \, .$$

This equality follows from the Poisson summation formula with

$$P_y(x) = \frac{c_n y}{(|x|^2 + y^2)^{[n+1]/2}} \quad \text{and} \quad \widehat{P}_y(\xi) = e^{-2\pi y|\xi|}$$

as usual. We claim that, on the fundamental cube,

$$\widetilde{P}_y(x) = P_y(x) + (\text{some harmless stuff}) \, .$$

This follows since

$$\sum_{m \neq 0} P_y(x + m) \equiv \sum{}' P_y(x + m) \leq A \sum \frac{y}{(|m|+1)^{n+1}} \leq A'y \, .$$

This is the error term.

Now we have

(1) $\widetilde{P}_y(x) \geq 0$.
(2) $\int_{T^n} \widetilde{P}_y(x)\, dx = 1$.
(3) $\lim_{y \to 0} \int_{\substack{T^n \\ |x| < \epsilon}} \widetilde{P}_y(x)\, dx = 1$.

Thus facts about almost everywhere convergence can be reduced to facts about the classical kernel.

Now suppose that $f \sim \sum a_m e^{2\pi im \cdot x}$. We have the Riesz transforms

$$\widetilde{R}_k(f) \sim (-i) \sum{}' a_m \frac{m_k}{|m|} e^{2\pi im \cdot x} \, .$$

The motivation for this definition is that if $f \in L^2(\mathbb{R}^n)$, then

$$(R_k f)\widehat{\ }(\xi) = (-i) \frac{\xi_k}{|\xi|} \widehat{f}(\xi) \, .$$

Suppose that

$$u(x, y) = \sum a_m e^{-2\pi |m| y} e^{2\pi i m \cdot x} .$$

We will also call this u_0. Let

$$u_k(x, y) = (-i) {\sum}' a_m \frac{m_k}{|m|} e^{-2\pi |m| y} e^{2\pi i m \cdot x} .$$

Then $u_0, u_1, \ldots, u_n$ are all harmonic and

$$\sum \frac{\partial u_j}{\partial x_j} = 0 \ \ (\text{with } x_0 = y)$$

as well as

$$\frac{\partial u_j}{\partial x_k} = \frac{\partial u_k}{\partial x_j} .$$

Theorem 2.4.1 *Let* $1 < p < \infty$. *Then* $f \in L^p(T^n)$ *implies that*

$$\|\widetilde{R}_k(f)\|_{L^p} \le A_p \|f\|_{L^p} .$$

Now let

$$Q_y^{(k)}(x) = \frac{c_n(x_k)}{(|x|^2 + y^2)^{[n+1]/2}}$$

and

$$\widehat{Q}_y(\xi) = (-i) \frac{\xi_k}{|\xi|} e^{-2\pi |\xi| y} , \ \ y > 0 .$$

We thus define

$$\widetilde{Q}_y^k = (-i) \sum \frac{m_k}{|m|} e^{-2\pi |m| y} e^{2\pi i m \cdot x} .$$

Lemma 2.4.2 *Let* $y > 0$. *Then*

$$\widetilde{Q}_y^k(x) = Q_y^k(x) + \left({\sum_m}' Q_y^k(x + m) \right) + (\text{error}) .$$

One can also see that

$$\frac{\partial \widetilde{Q}_y^k}{\partial y} = \frac{\partial \widetilde{P}_y(x)}{\partial x_k} + (\text{error}).$$

Finally, one can show that if $f \in L^p(T^n)$, then

$$\|f * Q_y^k\|_{L^p(T^n)} \le A_p \|f\|_{L^p(T^n)},$$

with A_p independent of y.

FACT If M is a multiplier for L^p in the non-periodic case, then $f|_\Lambda$ is a multiplier, when this makes sense. More precisely,

Theorem 2.4.3 *Let $\Phi \in L^\infty(\mathbb{R}^n)$. Suppose that Φ is defined at each $m \in \Lambda$ and that*

$$\Phi(m) = \lim_{r \to 0} \frac{1}{m(B_r)} \int_{B_r} \Phi(m - t) \, dt.$$

[This condition holds, in particular, if Φ is continuous at each m.] Suppose that

$$T_\Phi : f \mapsto \left(\Phi(\xi) \widehat{f}(\xi) \right)^{\vee}$$

satisfies

$$\|T_\Phi f\|_{L^p(\mathbb{R}^n)} \le A \|f\|_{L^p(\mathbb{R}^n)} \ , \ f \in L^p \cap L^2.$$

If we define $T_{\widetilde{\Phi}}$ on $L^p(T^n)$ by

$$T_{\widetilde{\Phi}}(f) \sim \sum a_m \Phi(m) e^{2\pi i m \cdot x}$$

whenever $f \sim \sum a_m e^{2\pi i m \cdot x}$, then we have

$$\|T_{\widetilde{\Phi}}(f)\|_{L^p(T^n)} \le A \|f\|_{L^p(T^n)}$$

with the same constant A as above.

This theorem is due to K. de Leeuw (see [STW1]. We apply the theorem as follows.

Let

$$\Phi(x) = \begin{cases} \frac{i x_k}{|x|} & \text{if} \quad x \ne 0 \\ 0 & \text{if} \quad x = 0. \end{cases}$$

An application of de Leeuw's theorem yields that the periodic Riesz transforms are of type (p, p), $1 < p < \infty$, with norms less than or equal to those of the ordinary Riesz transforms. There is a converse to de Leeuw's theorem.

This yields that the norms are in fact the same. Marcinkiewicz proved the boundedness of periodic Riesz transforms using his multiplier theorem.

Now let $f \in H^p(T^n)$, $f \sim \sum a_m e^{2\pi i m \cdot x}$. Also let

$$u(x, y) = \sum a_m e^{-2\pi |m| y} e^{2\pi i m \cdot x}$$

and

$$u^*(x) = \sup_{(x', y') \in \Gamma(x)} |u(x', y')|.$$

We further set

$$[S(u)]^2(x) = \left(\int_{\Gamma(x)} |\nabla u(x', y')|^2 (y')^{1-n} \, dx' dy' \right).$$

Let $\varphi \in \mathcal{S}(\mathbb{R}^n)$, $\varphi_\epsilon = \epsilon^{-n} \varphi(x/\epsilon)$. Then, finally, let $\widetilde{\varphi}_\epsilon = \sum_m \varphi_\epsilon(x + m)$. One then has

$$\int f(x - y) \varphi_\epsilon(y) \, dy = \int_{T^n} f(x - y) \widetilde{\varphi}_\epsilon(y) \, dy$$

$$= f * \widetilde{\varphi}_\epsilon(x)$$

$$= \sum a_m \widehat{\varphi}(\epsilon m) e^{-2\pi i m \cdot x}.$$

In the penultimate inequality we have used the fact that f is periodic.

Theorem 2.4.4 *Let f be a given tempered distribution on T^n. Then the following are equivalent, for $0 < p < \infty$.*

(1) $u^*(f) \in L^p(T^n)$.
(2) $Su(x) \in L^p(T^n)$. That is, $a_0 + \|Su\|_{L^p(T^n)} \approx \|u^*\|_{L^p(T^n)}$.
(3) For all $\varphi \in \mathcal{S}$, $\sup_{\epsilon > 0} |f * \varphi_\epsilon(x)| \in L^p$.
(4) For some $\varphi \in \mathcal{S}$, $\int \varphi \, dx \neq 0$, $\sup_{\epsilon > 0} |f * \varphi_\epsilon(x)| \in L^p$.
(5) $1 < p < \infty$ implies $f \in L^p(T^n)$.
(6) $(n - 1)/n < p < \infty$, $u(x, y)$ as above, $u_k(x, y) = (-i) \sum a_m [m_k / |m|]$ $e^{-2\pi |m| y} e^{2\pi i m \cdot x}$, $F(x, y) = (u(x, y), u_1(x, y), u_2(x, y), \ldots, u_n(x, y))$, then $\sup_{y > 0} \int_{T^n} |F(x, y)|^p \, dx < \infty$.
(7) ——————

PROBLEM (PROBABLY EASY)

(1) Obtain this theorem as a *consequence* of the corresponding result for $H^p(\mathbb{R}^n)$.

(2) Make precise the following: $f \in H^p(T^n)$ if and only if whenever $\varphi \in C_c^\infty(\mathbb{R}^n)$, then $\varphi \cdot f \in H^p(\mathbb{R}^n)$.
   Caution: In case $p = 1$, $f \in H^1(\mathbb{R}^n)$ requires $\int_{\mathbb{R}^n} f(x)\, dx = 0$ since $f, R_j f \in L^1$.

(3) Summability of Fourier series: The results for L^p are quite hard, but there is some hope for H^p, Stein thinks.[3] If $f \sim a_m e^{2\pi i m \cdot x}$, let

$$S_R^\delta f(x) = \sum_{|m| < R} a_m \left(1 - \frac{|m|^2}{R^2}\right)^\delta e^{2\pi i m \cdot x}.$$

In a weak sense, $S_R^\delta f \to f$ as $R \to \infty$.

Question Does it work in a better way? For $\delta = 0$, we get the spherical partial sum operator, which fails in L^p, $p \neq 2$.

Summability a.e. for S_R^δ

$n = 1$	$n = 2$
$\delta = 0$, $p > 1$, yes (Carleson-Hunt)	$\delta = 0$, $p = 2$, unknown; false if $p < 2$
$\delta = 0$, $p = 1$, false (Kolmogorov)	$f \in L(\log L)^2$, yes, $\delta = (n-1)/2$
$\exists f \in H^1$ w/ Fourier series div. a.e.	$f \in H^1$??
$\delta = 1/p - 1$, $p < 1$, yes (Zygmund)	$0 < p < 1$, $\delta = n(1/p - 1/2) - 1/2$???

Example 2.4.5 Consider $F(z) = \sum a_n z^n$. Note that

$$\frac{1}{(1-z)^\gamma} = \sum A_n^\gamma z^n, \quad \text{where } A_n^\gamma \sim c_\gamma n^{\gamma - 1}.$$

For example,

$$\frac{1}{(1-z)^2} = \sum n z^{n-1} \in H^{1/2 - \epsilon} \quad \text{(here } \gamma = 2 \text{ of course)}.$$

This shows that Zygmund's result is sharp.

THE MARCINKIEWICZ MULTIPLIER THEOREM:

Let

$$\mathcal{M}_k = \{\Phi \in C^k(\mathbb{R}^n \setminus \{0\}) : |\Phi(x)| \leq A, |x|^\alpha |\Phi^{(\alpha)}(x)| \leq A\alpha, |\alpha| \leq k\}.$$

---

[3] To this author's knowledge, this problem is still open.

Let $\mathcal{M}_\infty = \cap \mathcal{M}_k$. Write $(T_\Phi f)^\wedge = \Phi \hat{f}$. Note that $\Phi \in \mathcal{M}_0$ implies that T_Φ is of type $(2, 2)$.

Now we present a classical theorem.

Theorem 2.4.6 *Let $\Phi \in \mathcal{M}_k$, $k > n/2$. Then trivially $T_\Phi : L^2 \to L^2$ boundedly. But also T_Φ maps $L^p \cap L^2$ to $L^p \cap L^2$ boundedly. That is,*

$$\|T_\Phi f\|_{L^p} \leq A_p \|f\|_{L^p} \,, \ 1 < p < \infty \ \text{for } f \in L^p \cap L^2 \,.$$

As an instance, let $\Phi(x) = \Omega(x)$ with Ω homogeneous of degree 0, C^k on the unit sphere ($\Phi(x) = x_j/|x|$, for example).

Theorem 2.4.7 *Let $\Phi \in \mathcal{M}_k$, $k > n/p$, $0 < p \leq 1$. Then T_Φ yields a bounded operator on H^p.*

This last result is only known in special instances. See [FES].

Theorem 2.4.8 *Let $\Phi \in \mathcal{M}_k$, $f \sim \sum a_m e^{2\pi i m \cdot x}$. Then the operator*

$$f \mapsto \sum{}' a_m \Phi(m) e^{2\pi i m \cdot x}$$

is bounded from H^p to H^p if $k > n/p$, $0 < p \leq 1$ and $k > n/2$ for $1 < p < \infty$.

Note It is sometimes interesting to think of

$$H^p(T^n) \overset{?}{=} H^p(T^1 \times T^1 \times \cdots \times T^1) \,.$$

Consider the polydisc $\mathbb{D}^n = D^1 \times D^2 \times \cdots \times D^n$, each $D^j = \{z : |z| < 1, \text{ all } j\}$. Suppose that F is holomorphic on $\mathbb{D}^n$, $F(z) = \sum_{\substack{m \geq 0 \\ \text{multi-index}}} a_m z^m$. Here

$$z^m \equiv z_1^{m_1} \cdot z_2^{m_2} \cdot \cdots \cdot z_n^{m_n} \,.$$

Also $m \geq 0$ means $m_j \geq 0$ for all j.

Definition 2.4.9 Let $F \in H^p(\mathbb{D}^n)$, $0 < p < \infty$. For $z_j = r_j e^{i\theta_j}$, $0 \leq r_j < 1$, we have

$$\sup_{\text{all } r_j < 1} \int_0^{2\pi} \cdots \int_0^{2\pi} |F(r_1 e^{i\theta_1}, \ldots, r_n e^{i\theta_n})|^p \, d\theta_1 \ldots d\theta_n = \|F\|_{H^p(\mathbb{D}^n)} < \infty \,.$$

Some of what follows is due to Bochner for $p = 1$ and to Zygmund for general p. We can do many proofs by reduction to one variable.

Now we construct real variable H^p. Let $\theta_j = 2\pi x_j$, $r_j = e^{-y_j}$, $y_j \geq 0$. Let

$$F(x_1, \ldots, x_n) \sim \sum_{m \geq 0} a_m e^{2\pi i m \cdot x} .$$

Then

$$F(x_1 + iy_1, \ldots, x_n + iy_n) \sim \sum a_m e^{-2\pi m \cdot y} e^{2\pi i m \cdot x} .$$

So

$$\sup_{y>0} \int_{T^n} |F(x_1 + iy_1, \ldots, x_n + iy_n)|^p \, dx = \|F\|_{H^p}^p .$$

Let us now discuss the definition. We are trying to define $\mathbb{R}H^p(T^1 \times \cdots \times T^1)$. Let f be a tempered distribution on T^n, $f \sim \sum a_m e^{2\pi i m \cdot x}$. We wish to normalize f so that $a_m = 0$ whenever some $m_j = 0$. In fact, we write f as

$$f = f_{\text{normalized}} + (\text{other stuff}) ,$$

where the "other stuff" consists of functions of $(n-1)$ variables. We work only in dimension $n = 2$ and proceed in the following way.

Let

$$f_1 \sim \sum_{m_2=0} a_{m_1 m_2} e^{2\pi i m \cdot x} ,$$

$$f_2 \sim \sum_{m_1=0} a_{m_1 m_2} e^{2\pi i m \cdot x} ,$$

$$f_{12} = a_{00} .$$

Then

$$f - f_1 - f_2 + f_{12} = f_{\text{normalized}} .$$

Let $\eta = (\eta_1, \eta_2, \ldots, \eta_n)$, each $\eta_j = \pm 1$.

Definition 2.4.10 Let f be normalized. Define $F_\eta(z)$, D_η^n as we did on the octant. Write

$$f(x_1, \ldots, x_n) = \sum_\eta F_\eta(x_1, \ldots, x_n) ,$$

where each $F_\eta \in H^p(\mathbb{D}_\eta^n)$. We let the $\mathbb{R}H^p$ norm of f be

$$\sum_\eta \|F_\eta\|_{H^p(\mathbb{D}_\eta^n)} .$$

Theorem 2.4.11 *Let $f \sim \sum_m a_m e^{2\pi i m \cdot x} \in \mathbb{R}H^p$. Let*

$$u(x, y) = \sum_m a_m e^{-2\pi(|m_1||y_1| + \cdots + |m_n||y_n|)} e^{2\pi i m \cdot x}, \quad y_j > 0.$$

Let

$$u^*(x) = \text{unrestricted maximal function over the cone } \Gamma$$

$$= \sup_{\substack{x'_j + iy'_j \in \Gamma(x_j) \\ j=1,\ldots n}} |u(x'_1 + iy'_1, \ldots, x'_n + iy'_n)|.$$

Then $u^* \in L^p(T^n)$ and

$$\|u^*\|_{L^p(T^n)} \leq C \|f\|_{\mathbb{R}H^p(T^1 \times \cdots \times T^1)}.$$

Also unrestricted limits of u exist almost everywhere on T^n.

MAIN UNSOLVED PROBLEM: This is the converse. Let f be a given tempered distribution on T^n, $u(x, y)$ the multiple Poisson integral of f as above. Suppose that u^* is the unrestricted maximal function as above, and suppose that $u^* \in L^p$. Is it true that $u \in \mathbb{R}H^p(T^n)$? [The answer is well known if $1 < p < \infty$.]

PARTIAL RESULT: Assume $n = 2$. Begin with a function in $L \log L$ of T^2, i.e.,

$$\int_{T^2} |f|(1 + \log^+ |f|) \, dx < \infty.$$

Let

$$f \sim \sum a_{m_1 m_2} e^{2\pi i (m_1 x_1 + m_2 x_2)}.$$

Let Φ_1, Φ_2 be two one-dimensional multipliers of Marcinkiewicz type. That is, suppose that

$$\left| \frac{d^\ell \Phi_j(x)}{dx^\ell} \right| \leq A|x|^{-\ell} \text{ away from 0 for } \ell = 0, 1, \ldots, M, \text{ some } M.$$

Let

$$Tf \sim \sum a_{m_1 m_2} \Phi_1(m_1) \Phi_2(m_2) e^{2\pi i (m_1 x_1 + m_2 x_2)}.$$

QUESTION Does

$$\lim \sum a_{m_1 m_2} \Phi_1(m_1) \Phi_2(m_2) e^{-2\pi(|m_1||y_1| + |m_2||y_2|)} e^{2\pi i (m_1 x_1 + m_2 x_2)}$$

exist almost everywhere restrictedly?

ANSWER Yes. Argument is due to Cotlar for $L(\log L)^2$ and to C. Fefferman for $L \log L$.

Sketch of Proof
First Step: If $f \sim \sum a_{m_1 m_2} e^{2\pi i (m_1 x_1 + m_2 x_2)} \in L \log L$, then we claim that $\hat{f} \in \mathbb{R} H^p(T^1 \times T^1)$, $p < 1$. For let us assume that f has been normalized. Write $f = f_1 + f_2 + f_3 + f_4$, where

$$f_1 = \sum_{\substack{m_1 > 0 \\ m_2 > 0}} a_{m_1 m_2} e^{2\pi i (m_1 x_1 + m_2 x_2)} .$$

Since the Hilbert transform takes $L \log L$ to L, we get that $f_1 \in H^p$ for $p < 1$ via multiple applications of the one-dimensional Hilbert transform.

Lemma 2.4.12 *Suppose that*

$$f(x_1, x_2) \sim \sum_{\substack{m_1 \geq 0 \\ m_2 \geq 0}} a_{m_1 m_2} e^{2\pi i (m_1 x_1 + m_2 x_2)} \in H^p(T^1 \times T^1) , \ 0 < p < \infty.$$

Suppose that Φ is a multiplier of Marcinkiewicz type on $\mathbb{R}^1$. Then

$$\sum \Phi(m_1) a_{m_1 m_2} e^{2\pi i m \cdot x} \in H^p(T^1 \times T^1) .$$

Proof Instead of considering F, instead consider F_ϵ given by

$$F_\epsilon(z_1, z_2) = F(x_1 + i y_1 + i\epsilon, x_2 + i y_2 + i\epsilon) ,$$

$$F_\epsilon(x_1, x_2) = \sum_{\substack{m_1 > 0 \\ m_2 > 0}} a_{m_1 m_2} e^{-\epsilon 2\pi |m_1|} e^{-\epsilon 2\pi |m_2|} e^{2\pi i m \cdot x} .$$

Define

$$G_\epsilon(x_1, x_2) = \sum_{\substack{m_1 > 0 \\ m_2 > 0}} \Phi(m_1) a_{m_1 m_2} e^{-2\pi (|m_1| + |m_2|)} e^{2\pi i m \cdot x} .$$

Now fix ϵ, z_2 and consider the function

$$z_1 \mapsto F_\epsilon(z_1, z_2) .$$

Call this function $H(z_1)$. This function is in H^p of one variable. So

Fig. 2.7 The unit ball in $\mathbb{C}^n$

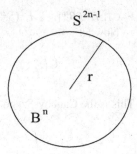

$$\int |G(x_1 + iy_1, x_2 + iy_2)|^p \, dx_1 \le C_p \|F\|_{H^p}^p$$

and

$$\int_{T^1} |G_\epsilon(x_1 + iy_1, x_2 + iy_2)|^p \, dx_1 \le C_p \int |F_\epsilon(x_1, x_2 + iy_2)|^p \, dx_1 .$$

Also

$$\iint_{T^1 \times T^1} |G_\epsilon(x_1 + iy_1, x_2 + iy_2)|^p \, dx_1 dx_2 \le C_p \iint_{T^1 \times T^1} |F_\epsilon(x_1, x_2 + iy_2)|^p \, dx_1 dx_2$$

$$\le C_p \iint_{T^1 \times T^1} |F(x_1 + i\epsilon, x_2 + iy_2 + i\epsilon)|^p \, dx_1 dx_2$$

$$\le C_p \|F\|_{H^p(T^2)} .$$

But

$$G_\epsilon(x_1 + iy_1, x_2 + iy_2) = G(x_1 + iy_1 + i\epsilon, x_2 + iy_2 + i\epsilon).$$

Thus letting $\epsilon \to 0$ gives the inequality for G.   □

We may now apply the lemma one variable at a time to prove that $f \in H^p$, $p < 1$.

QUESTION What are the analogues for the Riesz transforms on the ball in $\mathbb{C}^n$? What is the real variable analogue to the complex theory (the complex theory is known). See [CHG]. Now consider the Fig. 2.7.

The unit ball in $\mathbb{C}^n$ is $\{z = (z_1, \ldots, z_n) : |z| < 1, z_j \in \mathbb{C}\}$. Then

$$H^p(B^n) = \{F \text{ holomorphic on } B^n : \sup_{r<1} \left(\int_{S^{2n-1}} |F(rz)|^p \, d\sigma(z) \right)^{1/p} \equiv \|F\|_{H^p} < \infty\}.$$

Here $d\sigma$ is rotationally invariant area measure on $\partial B = S^{2n-1}$.

FACT: $H^2(B^n) \hookrightarrow L^2(S^{2n-1})$.

Now let

$$C(z, \zeta) = \frac{c_n}{(1 - z \cdot \overline{\zeta})^n} \ , \ z \cdot \overline{\zeta} \equiv z_1 \overline{\zeta}_1 + \cdots + z_n \overline{\zeta}_n .$$

This is the Cauchy-Szegő kernel for B^n. We have a map

$$f \mapsto F(z) = \int_{S^{2n-1}} C(z, \zeta) f(\zeta) \, d\sigma(\zeta) \ , \ c_n = \frac{(n-1)!}{2\pi^n} \ , \ f \in L^2(S^{2n-1}) .$$

This kernel was originally computed by Hua.

The mapping $f \to$ (boundary values of F) is an interesting projection operator to boundary values of holomorphic functions. Let

$$\mathcal{P}(z, \zeta) = \frac{|C(z, \zeta)|^2}{C(\zeta, \zeta)}$$

be the Poisson-Szegő kernel. Define

$$u(z) = \int_{S^{2n-1}} \mathcal{P}(z, \zeta) f(\zeta) \, d\sigma(\zeta) \ , \ f \in L^2(S^{2n-1}) .$$

There is a metric on B^n which is invariant under linear fractional transformations. In $\mathbb{C}^1$ it is the Poincaré or Lobachevsky metric given by

$$ds^2 = \frac{dx^2 + dy^2}{(1 - |z|^2)^2} .$$

Also, on B^1 the

$$\text{ordinary Laplacian} = \text{div grad}$$

and the Laplace–Beltrami operator (i.e., the operator that is invariant in the metric) coincide.

On B^n the metric is the Bergman metric arising from the Bergman reproducing kernel. The Laplacian and the Laplace–Beltrami operators definitely do *not* coincide.

From the form of the Cauchy kernel on B^n, we see that the fundamental singularity to be understood is $(1 - z \cdot \overline{\zeta})^n$. As $z \to \zeta \in \partial B^n$, how does this singularity behave?

Let $\zeta = (1, 0, \ldots, 0)$ and $z = (z_1, z_2, \ldots, z_n)$. Then

$$|1 - z \cdot \overline{\zeta}|^2 = |1 - z_1|^2 = (1 - x_1)^2 + (y_1)^2 . \tag{$*$}$$

A natural coordinate system on the surface cap is $(y_1, z_2, \ldots, z_n)$. So $(*)$ becomes

$$\sim [(1 - x_1)(1 + x_1)]^2 + (y_1)^2 = (1 - x_1^2)^2 + (y_1)^2 \qquad (\star)$$

since $x_1 \sim 1$ near ζ.

But

$$x_1^2 + y_1^2 + \sum_{j=2}^{n} |z_j|^2 = 1$$

so

$$(\star) \sim (y_1^2 + |z_2|^2 + \cdot + |z_n|^2)^2 + y_1^2 \approx (|z_2|^2 + \cdots + |z_n|^2)^2 + y_1^2.$$

So the behavior is different in the complex normal (y_2) direction than in the complex tangential $(z_2, \ldots, z_n)$ directions.

This leads naturally to the study of a $(2n-2)$-dimensional distinguished subspace of the real tangent space to ∂B^n at ζ.

Chapter 3
Background on H^p Spaces

3.1 Where Did H^p Spaces Get Started?

We define

$$H^\infty(\text{unit disc}) \equiv \{f \text{ holomorphic on the unit disc and bounded}\}.$$

In the early twentieth century, the theory of the Lebesgue integral made this theory accessible. The culmination of this development was in 1903 (Lebesgue). In 1906 we have

Theorem 3.1.1 (P. Fatou) *If F is holomorphic and bounded on the disc D, then F has radial limits almost everywhere in ∂D. Also F is uniquely determined by these boundary values.*

In the 1910s and 1920s, this subject was studied by F. Riesz and M. Riesz. Let

$$H^2(D) = \{F \text{ holomorphic on the disc with } \sup_{0<r<1} \int_0^{2\pi} |F(re^{i\theta})|^2 \, d\theta \equiv \|F\|_{H^2}^2 < \infty\}.$$

Write

$$F(z) = \sum_{n=0}^{\infty} a_n z^n.$$

Then

$$F \in H^2 \text{ if and only if } \left(\sum |a_n|^2\right)^{1/2} < \infty$$

© The Author(s), under exclusive license to Springer Nature Switzerland AG 2023
S. G. Krantz, *The E. M. Stein Lectures on Hardy Spaces*, Lecture Notes in
Mathematics 2326, https://doi.org/10.1007/978-3-031-21952-8_3

and $\sum a_n e^{in\theta} \sim f(e^{i\theta})$, the boundary function. Also $F(re^{i\theta}) \rightarrow f(e^{i\theta})$ as $r \rightarrow 1$ almost everywhere and in L^2. Note that $H^\infty \subseteq H^2$ trivially.

Now consider H^p, $0 < p < \infty$, given by

$$H^p(D) = \{F \text{ holomorphic on the disc with } \sup_{0<r<1} \left(\int_0^{2\pi} |F(re^{i\theta})|^p \, d\theta \right)^{1/p} < \infty\}.$$

Recall the L^p norm on ∂D with Lebesgue measure:

$$\left\{ f : \left(\int_0^{2\pi} |f(e^{i\theta})|^p \, d\theta \right)^{1/p} < \infty \right\}.$$

This is uninteresting for $p < 1$ since it no longer gives a norm. One can still define a metric, however, by

$$\rho_p(f, g) = \int_0^{2\pi} |f(\theta) - g(\theta)|^p \, d\theta.$$

There are no continuous linear functionals on this metric space. For, if

$$\varphi : L^p \rightarrow \mathbb{C} \ , \ p < 1 \ ,$$

satisfies

$$|\varphi(f) - \varphi(g)| \leq C\rho_p(f, g) ,$$

then

$$|\varphi(f)| \leq C\rho_p(f, 0) .$$

Hence, for any $\delta > 0$,

$$|\varphi(\delta f)| \leq C\rho_p(\delta f, 0)$$

so that

$$\delta|\varphi(f)| \leq \delta^p \cdot C\rho_p(f).$$

Leaving f fixed and letting $\delta \rightarrow +\infty$ gives a contradiction unless $\varphi(f) \equiv 0$ for all f.

Theorem 3.1.2 (M. Riesz, 1927) $H^p \sim L^p$, that is, the Hilbert transform is bounded on L^p, $1 < p < \infty$.

This theorem is false when $p = 1$. For $p < 1$, H^p functions have boundary values. We use Blaschke products to first isolate the zeros of the function, then look at $\widetilde{f}^{p/2} \in H^2$ where $f = \widetilde{f} \cdot b$ and b is a Blaschke product.

One could generalize all this to $\mathbb{R}^n$ via

(a) Fourier series
(b) The Hilbert transform
(c) n-tuples of harmonic functions which satisfy Cauchy–Riemann equations.

The following table illustrates the possibilities.

H^p in one complex variable	Possible real variable generalizations to $\mathbb{R}^n$
Holomorphic functions $\leftrightarrow$ harmonic functions satisfying Cauchy–Riemann equations	Generalizations of the Cauchy–Riemann equations
Hilbert transform	Singular integrals pseudo-differential operators Mikhlin–Calderón–Zygmund-Tricomi-Giroud
Simple interpretation in terms of Fourier transform	Simple formulas for Fourier transforms of kernels of operators of the above type
Real variable meaning of H^p, $p \leq 1$??	$H^1, H^p, p < 1$. Singular integral operators not bounded on L^1, but H^1 is the natural substitute. It is the natural space which is closed under all singular integral operators.

$$H^p \text{ theory in one complex variable} \begin{cases} \text{real variable theory in } \mathbb{R}^n \\ H^\infty \text{ as a Banach algebra} \\ \text{several complex variables} \end{cases}$$

Remark In one complex variable, there is one domain. In several complex variables there are a great many. For example,

(a) The unit ball in $\mathbb{C}^n$ (smooth boundary, strictly pseudoconvex).
(b) The polydisc—H^p for the polydisc is very different from that for the unit ball. One has a topological boundary which is strictly larger than the "distinguished boundary."

(c) The bounded, symmetric domains of Cartan. For example, $\{n \times n$ complex matrices (z_{ij}) with $zz^* < I\}$.

3.2 Hardy Spaces in $\mathbb{C}^1$

References: [ZYG, HOF, DUR, RUD2].

Preliminaries: Fourier series on $[0, 2\pi]$.

Define

$$L^p((0, 2\pi), d\theta) = \left\{ f : \|f\|_p = \left(\int_0^{2\pi} |f(\theta)|^p \, d\theta \right)^{1/p} \text{ is finite} \right\}, \quad 1 \le p < \infty,$$

$$\|f\|_\infty = \text{essential supremum } |f|.$$

Note that

$$L^\infty \subseteq L^p \subseteq L^1.$$

If $f \in L^1$, we say that $f \sim \sum_{n=-\infty}^{\infty} a_n e^{in\theta}$ if

$$a_n = \frac{1}{2\pi} \int_0^{2\pi} f(\theta) e^{-in\theta} \, d\theta.$$

(A) Convolution: If $f, g \in L^1$, then let

$$f * g(\theta) = \frac{1}{2\pi} \int_0^{2\pi} f(\theta - \varphi) g(\varphi) \, d\varphi.$$

Then $f * g \in L^1$ and $\|f * g\|_{L^1} \le (1/(2\pi)) \|f\|_{L^1} \|g\|_{L^1}$. If $f \sim \sum a_n e^{in\theta}$, $g \sim \sum b_n e^{in\theta}$, then $f * g \sim \sum a_n b_n e^{in\theta}$.

(B) Completeness: Let $f \in L^1$. Suppose that $a_n(f) = 0$ for all n. Then $f = 0$ almost everywhere.

Proof By hypothesis, $\int Pf = 0$ for any trigonometric polynomial P. So $\int gf = 0$ for any continuous function g on the circle. Hence $f = 0$ almost everywhere.

(C) Parseval theorem: If $f \in L^2$, then $\sum |a_n|^2 < \infty$ and

$$\frac{1}{2\pi} \int_0^{2\pi} |f(\theta)|^2 \, d\theta = \sum |a_n|^2.$$

[This implies that $S_n f \to f$ in L^2 as $n \to \infty$.]

Fig. 3.1 The mean value property

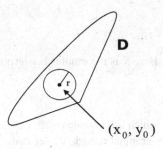

(D) Suppose that $f \in L^1$ and $\sum |a_n|^2 < \infty$. Then $\sum a_n e^{in\theta} = f(\theta)$ almost everywhere.

Proof Both sides have the same Fourier coefficients.

(E) If $f \in L^1$, then $|a_n| \leq (1/(2\pi))\|f\|_{L^1}$ and $a_n \to 0$ as $|n| \to \infty$.

Proof Write $f = f_1 + f_2$, where $f_1 \in L^2$ and $\|f_2\|_{L^1} < \epsilon$. Then $\lim_{|n| \to \infty} |a_n(f)| < \epsilon$ for any $\epsilon > 0$.

Harmonic Functions On $\mathbb{R}^n$,

$$\Delta = \sum_{j=1}^n \frac{\partial^2}{\partial x_j^2} \qquad \left(\Delta = \frac{\partial^2}{\partial x^2} + \frac{\partial^2}{\partial y^2} \text{ on } \mathbb{R}^2 \right).$$

If $\Omega \subseteq \mathbb{R}^n$ is open, then $u \in C^2(\Omega)$ is harmonic if $\Delta u = 0$.

Green's Theorem Let $\mathcal{R} \subseteq \mathbb{R}^n$ be a region (connected, open, bounded, smooth boundary). If u, v are C^2 on a neighborhood of $\overline{\mathcal{R}}$, then

$$\int_{\mathcal{R}} (u \Delta v - v \Delta u) \, dV(x) = \int_{\partial \mathcal{R}} u \frac{\partial v}{\partial \eta} - v \frac{\partial u}{\partial \eta} \, d\sigma .$$

Here ν is the unit outward normal vector and $d\sigma$ is boundary area measure (Hausdorff measure).

Mean Value Property Consider the figure. Suppose that u is a harmonic function on the domain $\mathcal{D}$. Assume that $B((x_0, y_0), r) \subseteq \mathcal{D}$. Then

$$u(x_0, y_0) = \frac{1}{2\pi} \int_0^{2\pi} u(x_0 + r \cos\theta, y_0 + r \sin\theta) \, d\theta .$$

Obviously this formulation is the two-dimensional version, but the result is true in any dimension (Fig. 3.1).

Proof We have, from Green's theorem with $u = u$ and $v \equiv 1$, that

$$\int_{\partial \mathcal{R}} \frac{\partial u}{\partial \eta} \, d\sigma = 0.$$

Here $\mathcal{R}$ is the annulus between the spheres with center (x_0, y_0) and radii ϵ and r. Let

$$v(x, y) = \log |(x, y) - (x_0, y_0)|,$$

which is harmonic on $\mathbb{R}^2 \setminus \{(x_0, y_0)\}$. Then, applying Green's formula to u, v and using our initial remark, we see that

$$\int_{|(x,y)-(x_0,y_0)|=\epsilon} u \frac{\partial v}{\partial \eta} \, d\sigma = \int_{|(x,y)-(x_0,y_0)|=r} u \frac{\partial v}{\partial \eta} \, d\sigma.$$

Letting $\epsilon \to 0$ gives the result. □

Converse to the Mean Value Property If $u \in C^2(\mathcal{R})$ and u satisfies the mean value property at each point in $\mathcal{R}$ for all sufficiently small r, then u is harmonic on $\mathcal{R}$. See the figure.

Proof Write

$$u(x_0, y_0) = \frac{1}{2\pi} \int_0^{2\pi} u(x_0 + r \cos \theta, y_0 + r \sin \theta) \, d\theta \ , \ r \text{ small}.$$

Think of this as a function of r, and differentiate twice with respect to r. Thus we have

$$\frac{1}{2\pi} \int_0^{2\pi} \left(u_{xx} \cdot (\cos \theta)^2 + u_{yy} \cdot (\sin \theta)^2 + u_{xy} \cdot \cos \theta \sin \theta \right) d\theta = 0.$$

Evaluating at $r = 0$ yields

$$C \left(u_{xx} + u_{yy} \right) (x_0, y_0) = 0.$$

But $C \neq 0$ so that ends the proof. □

The Maximum Principle Suppose that $\mathcal{D} \subseteq \mathbb{R}^n$ is open, connected, bounded and that u is harmonic on $\mathcal{D}$ and continuous on $\overline{\mathcal{D}}$. Then

$$|u(z)| \le \sup_{\zeta \in \partial \mathcal{D}} |u(\zeta)| \text{ for all } z \in \mathcal{D}.$$

The Strong Maximum Principle Suppose that $\mathcal{D} \subseteq \mathbb{R}^n$ is open, u is harmonic on $\mathcal{D}$, and u is non-constant. Then, if $z \in \mathcal{D}$ one has

$$|u(z)| < \limsup_{\substack{\zeta \in \mathcal{D} \\ \zeta \to \partial \mathcal{D}}} |u(\zeta)|.$$

The Dirichlet Problem Let $\mathcal{R} \subseteq \mathbb{R}^n$ be a region, and suppose that f is a given continuous function on $\partial\mathcal{R} = \overline{\mathcal{R}} \setminus \mathcal{R}$. The problem is to find a function u satisfying the following:

  (i) u is harmonic in $\mathcal{R}$.
 (ii) u is continuous on $\overline{\mathcal{R}}$.
(iii) $u|_{\partial\mathcal{R}} = f$.

Solution of the Dirichlet Problem on the Disc
Let $f(\theta)$ be a given continuous 2π-periodic function, say $f(\theta) =\sim \sum a_n e^{in\theta}$. Note that, for the function $e^{in\theta}$, the solution to the Dirichlet problem is $r^n e^{in\theta} = z^n$, $n \geq 0$. When $n < 0$, the solution is $\overline{z^{-n}} = r^{|n|} e^{in\theta}$. So one guesses that $u = \sum_{n=-\infty}^{\infty} a_n r^{|n|} e^{in\theta}$ solves the problem in general. For $r < 1$, this function is well defined and harmonic.

We are thus let to define the *Poisson kernel*

$$P_r(\theta) = \sum_{n=-\infty}^{\infty} r^{|n|} e^{in\theta}.$$

[Note that, in [ZYG], there is a factor of $1/2$, but this is not important.] Write

$$u(r, \theta) = P_r * f(\theta)$$
$$= \frac{1}{2\pi} \int_{-\pi}^{\pi} P_r(\theta - \varphi) f(\varphi)\, d\varphi$$
$$= \frac{1}{2\pi} \int_{-\pi}^{\pi} P_r(\varphi) f(\theta - \varphi)\, d\varphi.$$

One can check that, for $r < 1$ and fixed,

$$u(r, \theta) \sim \sum_{n=-\infty}^{\infty} a_n r^{|n|} e^{in\theta}.$$

Since the a_ns are bounded, this works.
By straightforward algebra,

$$P_r(\theta) = \frac{1 - r^2}{1 - 2r \cos\theta + r^2}.$$

We will show that $u(r, \theta) = PI(f)$ is harmonic on the disc and $u(r, \theta) \to f(\theta)$ as $r \to 1$, uniformly when f is continuous. [n.b. In what follows, we will endeavor to write S for the circle, D for the disc, $h(D)$ for the harmonic functions on the disc, $H(D)$ the holomorphic functions, $C^j(D)$, $L^p(S)$, etc. the usual function spaces.]
Observe that

Fig. 3.2 The Poisson kernel

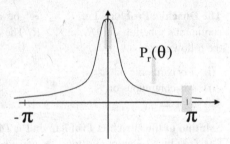

$$\frac{1}{2\pi} \int_0^{2\pi} P_r(\theta)\, d\theta = 1 \quad \text{for all } r\,.$$

Also $P_r(\theta) \geq 0$. This positivity follows on *a priori* grounds since harmonic functions satisfy the maximum principle. The figure shows a graph of $P_r(\theta)$ (Fig. 3.2).

Let $\delta = 1 - r$. Then, trivially,

$$P_r(\theta) = \frac{(1-r)(1+r)}{(1-r)^2 + 2r(1-\cos\theta)} \leq \frac{2\delta}{\delta^2 + \theta^2} \quad \text{for } r \geq \frac{1}{2}\,.$$

If $r < 1/2$, then $P_r(\theta) \leq (1+r)/(1-r) \leq 3$.

This is an important estimate which arises always and reflects the fundamental nature of the theory. [Estimates of a different (non-isotropic) nature, which arise in several complex variables, give rise to a different notion of convergence.]

Now

$$u(r,\theta) - f(\theta) = \frac{1}{2\pi} \int_{-\pi}^{\pi} P_r(\varphi)\, [f(\theta - \varphi)]\, d\varphi - f(\theta)$$

$$= \frac{1}{2\pi} \int_{-\pi}^{\pi} P_r(\varphi)\, [f(\theta - \varphi) - f(\theta)]\, d\varphi\,. \qquad (*)$$

Let $\omega_\infty(\varphi) = \sup_\theta |f(\theta - \varphi) - f(\theta)|$. Since f is uniformly continuous on S, $\omega_\infty(\varphi) < \infty$ for all φ. From $(*)$ we see that

$$\|u(r,\theta) - f(\theta)\|_\infty \leq \frac{1}{2\pi} \int_{-\pi}^{\pi} P_r(\varphi)\omega_\infty(\varphi)\, d\varphi\,.$$

We wish to make $\|u(r, \theta - f(\theta)\|_\infty < \epsilon$ for r sufficiently near to 1.

Pick $\eta > 0$ such that $\omega_\infty(\varphi) < \epsilon/2$ for $|\varphi| < \eta$. Then

$$\frac{1}{2\pi} \int_{-\pi}^{\pi} P_r(\varphi)\omega_\infty(\varphi)\, d\varphi \leq \frac{1}{2\pi} \int_{|\varphi| < \eta} + \frac{1}{2\pi} \int_{|\varphi| \geq \eta} \equiv T_1 + T_2\,.$$

Now

$$T_1 \leq \frac{\epsilon}{2} \cdot \frac{1}{2\pi} \int_{-\pi}^{\pi} P_r(\varphi)\, d\varphi = \frac{\epsilon}{2},$$

and

$$T_2 \leq \frac{1}{2\pi} \int_{|\varphi| \geq \eta} 2\|f\|_{L^\infty} \cdot \frac{\delta}{\delta^2 + \varphi^2}\, d\varphi \leq 2\|f\|_{L^\infty} \cdot \frac{\delta}{\eta^2}.$$

For δ sufficiently small (i.e., r sufficiently near 1), this last expression is less than $\epsilon/2$. So

$$\|u(r,\theta) - f(\theta)\|_\infty < \epsilon.$$

Now we know that we can recover f from the series $\sum a_n e^{in\theta}$ by Abel summability.

Theorem 3.2.1 (Fatou) *If $F \in H(D)$ and bounded, then for almost every θ, $\lim_{r \to 1^-} F(re^{i\theta})$ exists.*

We will denote the limit function by $F(e^{i\theta})$.

Lemma 3.2.2 *A necessary and sufficient condition that $u(r,\theta) \in h(D)$ be bounded is that*

$$u(r,\theta) = P_r * f \quad \text{for some } f \in L^\infty(S).$$

Proof If $f \in L^\infty$, then $\|P_r * f\|_{L^\infty} \leq \|P_r\|_1 \cdot \|f\|_\infty \leq \|f\|_\infty$. Also $P_r * f(\theta) \in h(D)$ by the arguments given before.

Conversely, for any $0 < \rho < 1$, let $f_\rho(\theta) = u(\rho, \theta)$. Since u is bounded, the f_ρ are uniformly in $L^\infty(S)$. Now

$$u(r\rho, \theta) = P_r * f_\rho(\theta) = \int P_r(\theta - \varphi) f_\rho(\varphi)\, d\varphi.$$

Let $\rho = 1 - 1/n$. By weak-$*$ compactness, there exists $\{n_k\}$ such that f_{ρ_k} converges weak-$*$ (here $\rho_k = 1 - 1/n_k$). Call the weak-$*$ limit f. So, for fixed r, θ,

$$P_r * f_{1-1/n_k}(\theta) \to P_r * f(\theta)$$

as $k \to \infty$. Also

$$P_r * f_{1-1/n_k}(\theta) = u(r(1 - 1/n_k), \theta) \to u(r, \theta)$$

by continuity.

As a result, $u(r,\theta) = P_r * f(\theta)$. $\qquad\square$

Fig. 3.3 An interval centered
at θ_0

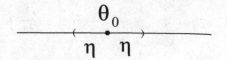

Definition 3.2.3 Suppose that $f \in L^1_{\text{loc}}(\mathbb{R}^1)$. We say that $x \in$ Lebesgue set of f if

$$\frac{1}{2h} \int_{B(0,h)} |f(x - y) - f(x)| \, dy \to 0 \text{ as } h \to 0.$$

Lemma 3.2.4 *Let $f \in L^1_{\text{loc}}(\mathbb{R}^1)$. Then almost every $x \in \mathbb{R}^1$ is in the Lebesgue set of f.*

Proof We will prove this later in greater generality.   □

To conclude the proof of Fatou's theorem, we have $u(r, \theta)$ bounded and harmonic if $u \in H^\infty$, so $u = P_r * f$ for some $f \in L^\infty(S)$.

Now assume only $f \in L^1(S)$. We claim that $u(r, \theta) \to f(\theta)$ as $r \to 1$ for each θ in the Lebesgue set of f. For let θ_0 be in the Lebesgue set. Consider an interval of diameter 2η centered at θ_0. See Fig. 3.3.

Write $f = f_1 + f_2$, where

$$f_1 = \begin{cases} f & \text{on the interval} \\ 0 & \text{elsewhere.} \end{cases}$$

Now

$$u(r, \theta_0) - f(\theta_0) = u_1(r, \theta_0) - f_1(\theta_0) + u_2(r, \theta_0) - f_2(\theta_0).$$

Here $u_1 = PI\, f_1$ and $u_2 = PI\, f_2$ so that $u = u_1 + u_2$.

We have that $\lim_{r \to 1} u_2(r, \theta_0) = 0$ since

$$|u_2(r, \theta_0)| = \left| \frac{1}{2\pi} \int_{-\pi}^{\pi} P_r(\theta_0 - \varphi) f(\varphi) \, d\varphi \right|$$

$$\leq \frac{1}{2\pi} \int_{|\varphi| \geq \eta} P_r(\theta_0 - \varphi) |f(\varphi)| \, d\varphi$$

$$\leq \frac{1}{2\pi} \left(\sup_{|\varphi| \geq \eta} P_r(\varphi) \right) \cdot \|f\|_{L^1} \to 0$$

as $r \to 1$.

Also, of course, $f_2(\theta_0) = 0$. So we must estimate

$$|u_1(r,\theta) - f_1(\theta)| = \left| \int P(r,\varphi)[f_1(\theta - \varphi) - f_1(\theta)] \, d\varphi \right|$$

$$\leq \left| \int_{|\varphi| \leq \eta} \right| + \left| \int_{|\varphi| > \eta} \right|$$

$$\equiv T_1 + T_2.$$

As usual, $T_2 \to 0$. In order to estimate T_1, we reformulate the statement "θ_0 is in the Lebesgue set" as

$$\frac{1}{2h} \int_{|\varphi| \leq h} |f(\theta - \varphi) - f(\theta)| \, d\varphi < \epsilon(\eta) \;, \; \epsilon(\eta) \to 0 \text{ as } \eta \to 0.$$

Now

$$T_1 \leq C \int_{|\varphi| \leq \eta} \frac{\delta}{\delta^2 + \varphi^2} \cdot |f(\theta - \varphi) - f(\theta)| \, d\varphi$$

$$\leq C \int_{|\varphi| \leq \delta} + C \int_{|\varphi| > \delta}$$

$$\leq C \int_{|\varphi| \leq \delta} + C \sum_{\substack{k=0 \\ k \leq \log_2(\eta/\delta)}}^{\infty} \int_{\substack{2^k \delta \leq |\varphi| \leq 2^{k+1} \delta}} \cdot$$

Recall that δ denotes $1 - r$. Suppose that $\delta < \eta/10$. Now

$$\int_{|\varphi| \leq \delta} \frac{\delta}{\delta^2 + \varphi^2} \cdot |f(\theta - \varphi) - f(\theta)| \, d\varphi \leq \frac{1}{\delta} \int_{|\varphi| \leq \delta} |f(\theta - \varphi) - f(\theta)| \, d\varphi = \epsilon(\eta).$$

Now consider the kth term in the sum: When $2^k \delta \leq |\varphi| \leq 2^{k+1} \delta$, we have

$$P_r(\varphi) \leq \frac{C\delta}{\delta^2 + \varphi^2} \leq \frac{C\delta}{(2^k \delta)^2} = C\delta^{-1} \cdot 2^{-2k}.$$

So the kth term is

$$\leq C 2^{-2k} \delta^{-1} \int_{|\varphi| \leq 2^{k+1} \delta} |f(\theta - \varphi) - f(\theta)| \, d\varphi \leq C 2^{-k} \cdot \epsilon(\eta)$$

since $2^{k+1} \delta \leq \eta$ by design. So the sum is $\leq C \cdot \epsilon(\eta)$ and altogether

$$T_1 \leq K \cdot \epsilon(\eta),$$

which is what we desired. $\square$

Remarks About Fatou's Theorem : If $u \in H(D) \cap L^\infty(D)$ or even if $u \in h(D) \cap L^\infty(D).$, then we have

Fig. 3.4 The nontangential
approach region Γ_α

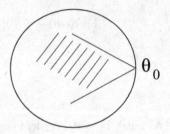

(1) $u(r, \theta) = P_r * f$ with $f \in L^\infty(S)$.
(2) u has radial limits.

The same proof, with a few modifications, shows that nontangential limits exist. To wit,
we let

$$\Gamma_\alpha(\theta_0) = \{(r, \theta) \in D : |\theta - \theta_0| < (1 + \alpha)(1 - r)\} \text{ for } \alpha > 0.$$

See the figure. We show that $f \in L^1$ and $u = PI(f)$ implies

$$\lim_{\substack{(r,\varphi) \in \Gamma_\alpha(\theta_0) \\ r \to 1}} = f(\theta_0)$$

for any θ_0 in the Lebesgue set of f with α fixed (Fig. 3.4).

The key observation is that

$$P(r, \varphi) \le C_\alpha P(r, \theta_0) \le \frac{\delta}{\delta^2 + (\theta_0)^2}. \tag{$*$}$$

for $(r, \varphi) \in \Gamma_\alpha(\theta_0)$.

Proof of (*): We have

$$
\begin{aligned}
P(r, \varphi) &= \frac{1 - r^2}{1 - 2r \cos \theta + r^2} \\
&\le \frac{C\delta}{\delta^2 + \varphi^2} \\
&\le \frac{2(1 + \alpha)^2 C\delta}{2(1 + \alpha)^2 \delta^2 + \varphi^2} \\
&\le \frac{C_\alpha \delta}{2(1 + \alpha)^2 \delta^2 + \theta_0^2 - |\theta_0 - \varphi|^2}
\end{aligned}
$$

just because $\varphi^2 \ge (1/2)\theta_0^2 - |\theta_0 - \varphi|^2$. And this last is

$$\leq \frac{C_\alpha \delta}{2(1+\alpha^2)\delta^2 + \theta_0^2 - (1+\alpha)^2\delta^2} \leq \frac{C_\alpha'\delta}{\delta^2 + \theta_0^2}.$$

And that finishes the proof.   □

We also have that u is the Poisson integral of an f in $L^p(S)$, $1 < p \leq \infty$, if and only if

$$\sup_{r<1}\left(\int_0^{2\pi} |u(r,\theta)|^p\, d\theta\right)^{1/p} < \infty$$

and $u \in h(D)$. Additionally, $u(r,\theta)$ is the Poisson integral of a finite Radon measure μ if and only if

$$\sup_{r<1}\int_0^{2\pi} |u(r,\theta)|\, d\theta < \infty$$

and

$$u(r,\theta) = \frac{1}{2\pi}\int_0^{2\pi} P_r(\theta - \varphi)\, d\mu(\varphi).$$

Thus, if $\sup_{r<1}\int_0^{2\pi} |u(r,\theta)|\, d\theta < \infty$, then $\lim_{r\to 1} u(r,\theta)$ exists almost everywhere Lebesgue. The limit also exists in the nontangential sense.

To see why the last remark is true, note that if $d\mu$ is a Radon measure, then

$$d\mu = f(\theta)d\theta + d\nu,$$

where $f \in L^2(S)$ and $d\nu \perp dx$. We already know that $P_r * f$ behaves well. Also,

$$\frac{1}{2h}\int_{x_0-h}^{x_0+h} |d\nu| \to 0$$

for almost every x_0 so that $P_r * d\nu \to 0$ almost everywhere.

As an example, consider the Dirac measure δ. Then $P_r * \delta(\theta) = P_r(\theta) \to 0$ as $r \to 1$ except at $\theta = 0$. The same proof gives the H^p result: of course one must be careful in the case of $p = 1$ because of non-duality.

Remark 3.2.5 This is in the one-dimensional case. Let $u(r,\theta) = P_r f$ with $f \in L^1_{\text{loc}}$. Then, at every point in the Lebesgue set, i.e., at which

$$\frac{1}{2h}\int_{-h}^{h} |f(\theta - \varphi) - f(\theta)|\, d\theta \to 0,$$

we have that $u \to f$ radially. The following *weaker* condition suffices:

$$\frac{1}{2h} \int_{-h}^{h} f(\varphi)\, d\varphi = f(\theta). \qquad\qquad (*)$$

[The proof requires integration by parts—a truly one-dimensional argument.] For nontangential convergence, however, $(*)$ does not suffice.

3.3 The Hardy–Littlewood Maximal Function

Let

$$Mf(\theta) = \sup_{0 < h < \pi} \frac{1}{2h} \int_{-h}^{h} |f(\theta - \varphi)|\, d\varphi.$$

This is the classical maximal function of Hardy and Littlewood. Now we formulate the notion in a much more general setting.

Let X be a locally compact space and let dx be a σ-finite measure on the space X. Let

$$m(E) = \int \chi_E\, dx.$$

Suppose that, for each $x \in X$ and $r > 0$, we are given

$$B(x, r) = \text{"the ball centered at } x \text{ of radius } r\text{"}.$$

Further suppose that the $B(x, r)$ satisfy

(1) Each $B(x, r)$ is open and bounded and $m(B(x, r)) > 0$.
(2) $B(x, r_1) \subseteq B(x, r_2)$ if $r_2 \geq r_1$.
(3) $m(B(x, 2r)) \leq C_1 m(B(x, r))$, some C_1 independent of x.
(4) If $B(x_1, r_1) \cap B(x_2, r_2) \neq \emptyset$ and $r_2 \geq r_1$, then $B(x_2, C_2 r_2) \supseteq B(x_1, r_1)$, some $C_2 \geq 1$ a fixed constant.

If $f \in L^p(X, dx)$, we set

$$Mf(x) = \sup_{r > 0} \frac{1}{m(B(x, r))} \int_{B(x,r)} |f(y)|\, dy.$$

Theorem 3.3.1 *We have that*

(1) *$Mf < \infty$ almost everywhere.*
(2) *$\|Mf\|_p \leq A_p \|f\|_p$, $1 < p \leq \infty$.*
(3) *For $p = 1$,*

$$m\{x : Mf(x) > \alpha\} \leq C \cdot \frac{\|f\|_{L^1}}{\alpha}.$$

Remark 3.3.2 One can see that **(3)** is best possible for L^1 by looking at $M\delta$, where δ is the Dirac measure, and recalling that δ is the weak-$*$ limit of an L^1 bounded sequence of L^1 functions.

In order to prove the theorem we need a covering lemma.

Lemma 3.3.3 *Suppose that*

$$E \subseteq \bigcup_{\substack{\text{finite} \\ \text{union}}} B(x_\alpha, r_\alpha).$$

Then there exists a subcollection of balls $B(x_1, r_1), B(x_2, r_2), \ldots B(x_n, r_n)$ *such that this subcollection is pairwise disjoint and also*

$$\sum_{j=1}^{n} m(B(x_j, r_j)) \geq C_3^{-1} \cdot m(E)$$

for some $C_3 > 0$.

Proof Pick $B(x_1, r_1)$ of maximal radius. Consider all balls which intersect $B(x_1, r_1)$. They are contained in $B(x_1, C_2 r_1)$. Now pick $B(x_2, r_2)$ disjoint from $B(x_1, r_1)$ and of maximal radius. Keep going. Since the original collection of balls is finite, the process stops.

Now

$$M(E) \leq m\left(\bigcup_j B(x_j, r_j)\right)$$

$$\leq \sum_{j=1}^{n} m(B(x_j, C_2 r_j))$$

$$\leq C_3 \cdot \sum_{j=1}^{n} m(B(x_j, r_j)).$$

In the last inequality we used property **(2)** of our space X finitely many times. □

Proof of the Theorem The proof is by interpolation of the trivial L^∞ result and the weak-type L^1 result which follows. [**Remark:** One has, for any g, $m\{x : |g(x)| > \alpha\} \leq \|g\|_{L^1}/\alpha$. This is Tchebycheff's inequality.]

Let $E_\alpha = \{x : Mf(x) > \alpha\}$. We need to show that

$$m(E_\alpha) \leq \frac{C_2}{\alpha} \|f\|_{L^1}.$$

Let $K \subseteq E_\alpha$ be compact. It suffices to show that

$$M(K) \leq \frac{C_2}{\alpha} \|f\|_{L^1}$$

by inner regularity. Let $x_0 \in K$. So $Mf(x_0) > \alpha$. Thus there exists an $r_0 > 0$ such that

$$\frac{1}{m(x_0, r_0)} \int_{B(x_0, r_0)} |f(y)| \, dy > \alpha.$$

Consider the set of all such $B(x_0, r_0)$ for $x_0 \in K$. By compactness, there exists a finite subcollection of these balls

$$B(x_1, r_1), B(x_2, r_2), \ldots, B(x_n, r_n)$$

which cover K (because K is compact). And for each $j = 1, \ldots, n$ we have

$$\int_{B_j} |f| \geq \alpha m(B_j).$$

Now we use the lemma to find a pairwise disjoint subcollection of the B_j, call it $B_{j_1}, B_{j_2}, \ldots, B_{j_s}$, which satisfies

$$\sum_{k=1}^{s} m(B_{j_k}) \geq C_3^{-1} m(K).$$

As a result,

$$m(K) \leq C_2 \sum_{k=1}^{s} m(B_{j_k})$$

$$\leq \frac{C_2}{\alpha} \sum_{k=1}^{s} \int_{B_{j_k}} |f|$$

$$\leq \frac{C_2}{\alpha} \|f\|_{L^1}$$

since the B_{j_k} are pairwise disjoint. This proves part **(3)** of the theorem.

Now we prove **(2)** using **(3)**. Suppose that $f \geq 0$. Let

$$\lambda_f(\alpha) = m\{x : f(x) > \alpha\}.$$

Call λ_f the *distribution function* of f. [**Fact:** It holds that

$$\int f(x)^p \, dx = - \int_0^\infty \alpha^p \, d\lambda_f(\alpha). \tag{$*$}$$

The proof of this fact is trivial for simple functions. The general case follows by approximation. It is also the case that $(*) = p \int_0^\infty \alpha^{p-1} \lambda_f(\alpha) \, d\alpha$.]

Claim Whenever $f \mapsto Mf$ is a sublinear map satisfying **(3)** and $|Mf| \le M(|f|)$ and also $M(\text{constant}) = \text{constant}$, then we can prove **(2)**. Now fix α. Let us define

$$f_1 = \begin{cases} f & \text{if} \quad f > \alpha/2 \\ 0 & \text{otherwise}. \end{cases}$$

Hence

$$|f| \le |f_1| + \frac{\alpha}{2}$$

and

$$Mf \le Mf_1 + M(\alpha/2) = Mf_1 + \frac{\alpha}{2}.$$

As a result,

$$Mf(x) > \alpha \Rightarrow Mf_1(x) > \frac{\alpha}{2}$$

and

$$m\{x : Mf(x) > \alpha\} \le m\{x : Mf_1(x) > \alpha/2\}. \qquad (\star)$$

Therefore, it suffices to estimate $m\{x : Mf_1(x) > \alpha/2\}$. But

$$(\star) \le \frac{2C_2}{\alpha} \int_{|f|>\alpha/2} |f| \, dx$$

by the L^1 result. Hence

$$m\{x : Mf(x) > \alpha\} \le \frac{2C_2}{\alpha} \int_{|f|>\alpha/2} |f| \, dx.$$

Therefore, we may calculate

$$\int |Mf(x)|^p \, dx = p \int_0^\infty \alpha^{p-1} \lambda_{Mf}(\alpha) \, d\alpha$$

$$\le p2C_2 \int_0^\infty \int_{|f|>\alpha/2} \alpha^{p-2} |f| \, dx d\alpha$$

$$\le p2C_2 \int_{\mathbb{R}^n} \int_0^{2|f|} \alpha^{p-2} |f| \, d\alpha dx$$

Fig. 3.5 Two intersecting balls in X

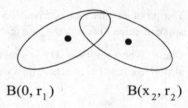

$$B(0, r_1) \qquad\qquad B(x_2, r_2)$$

$$\leq \frac{2C_2 p}{p-1} \int_{\mathbb{R}^n} |f|(2|f|)^{p-1}\, dx$$

$$= A_p^p \|f\|_p^p.$$

That is to say, $\|Mf\|_p \leq A_p \|f\|_p$. That ends the proof. □

Example 3.3.4 Let $X = \mathbb{R}^n$, $dx =$ Lebesgue measure. Let K be a convex, open, bounded, symmetric set. Define $B(0, r) = rK$ and $B(x, r) = x + B(0, r)$. We claim that $C_1 = 3^n$, $C_2 = 3$ in this case.

For if $B(x_1, r_1) \cap B(x_2, r_2) \neq \emptyset$, $r_1 \geq r_2$, then suppose without loss of generality that $x_1 = 0$. Refer to Fig. 3.5.

Suppose that $y \in B(0, r_1) \cap B(x_2, r_2)$ so that $y \in r_1 K$ and $y - x_2 \in r_2 K$ by definition of the balls. Let $z \in B(x_2, r_2)$ be arbitrary. Then $x_2 - z \in r_2 K$ so that

$$x_2 = (x_2 - y) + y \in r_2 K + r_1 K$$

and

$$z = (z - x_2) + x_2 \in r_2 K + x_2 \subseteq r_2 K + r_2 K + r_1 K \subseteq 3 r_1 K.$$

This shows that $C_2 = 3$. The claim for C_1 is trivial by dimensionality.

Example 3.3.5 Let $K = B(0, 1) = \{x \in \mathbb{R}^n : \sum_{j=1}^n |x_j|^2 < 1\}$. Let $r \circ x = (r^{a_1} x_1, \ldots, r^{a_n} x_n)$ with $a_1, \ldots, a_n > 0$. Define $B(x, r) = x + r \circ K$. We claim that, in this case,

$$C_1 \leq 3^{\sum a_j / \min a_j}.$$

(the first example is a corollary of this one.)

Remark 3.3.6 It is not abundantly clear that

$$\sup_{r>0} \frac{1}{m(B(x,r))} \int_{B(x,r)} |f(y)|\, dy$$

is measurable. But if

$$\chi(x, y, r) = \text{characteristic function of } \{y \in B(x, r)\},$$

and if we assume that $m(B(x, r))$ is continuous in r, then it is easy to check that $\chi(x, y, r)$ is product measurable and that the above supremum gives a measurable function.

Corollary 3.3.7 (Of the Maximal Theorem) *Recall that* $\|Mf\|_p \leq A_p\|f\|_p$, $1 < p \leq \infty$, *and also* $m\{x : |Mf(x)| > \alpha\} \leq (C/\alpha)\|f\|_1$. *Suppose that, for all* x, $\cap_r \overline{B(x, r)} = \{x\}$. *Let* $f \in L^p$, $1 \leq p \leq \infty$. *Then, for almost every* x,

$$\lim_{r \to 0} \frac{1}{m(B(x, r))} \int_{B(x,r)} f(y)\, dy = f(x) \text{ almost everywhere.}$$

Proof for $1 < p < \infty$:

Lemma 3.3.8 *Suppose that* A_r *is a sequence of operators such that if we let* $A_M f(x)$ *denote* $\sup_r |A_r f(x)|$, *then*

$$\|A_M f\|_p \leq C\|f\|_p\,.$$

Suppose also that $\lim_{r \to 0} A_r f \equiv A_0(f)$ *exists almost everywhere whenever* $f \in C_0$. *Then in fact* $A_0 f(x)$ *exists almost everywhere for all* $f \in L^p$. *[Observe, for instance, that*

$$\lim_{r \to 0} \frac{1}{B(x, r)} \int_{B(x,r)} f(y)\, dy = f(x)$$

for all x whenever $f \in C_0$.]

Proof of the Lemma Let $f \in L^p$. Write $f = f_1 + f_2$ with $f_2 \in C_0$ and $\|f_1\|_p \leq \delta$. Let

$$A_r f(x) = \frac{1}{m(B(x, r))} \int_{B(x,r)} f(y)\, dy\,.$$

Write

$$A_r f - f = (A_r f - f_1) + (A_r f_2 - f_2)\,.$$

Since the second term tends to 0 as $r \to 0$, it suffices to investigate $\limsup(A_r f_1 - f_1)$ almost everywhere.
   But

$$\limsup |A_r f_1 - f_1| \leq \sup |A_r f_1 - f_1|$$

$$\leq |Mf_1| + |f_1|$$

$$= F_1\,.$$

Of course

$$\|F_1\|_p \le \|Mf_1\|_p + \|f_1\|_p \le C\|f_1\|_p.$$

Now fix $\epsilon > 0$. By choosing δ sufficiently small, we may guarantee that $|F_1| \le \epsilon$ off of a set of measure ϵ. This is because

$$m\{x : |F_1| > \epsilon\} \le \frac{\|F_1\|_p^p}{\epsilon^p} \le \frac{C^p\|f_1\|_p^p}{\epsilon^p} \le \frac{C^p\delta^p}{\epsilon^p}.$$

But this says that

$$\limsup_{r \to 0} |A_r f_1 - f_1| \le \epsilon \text{ off a set of measure } \epsilon.$$

Therefore,

$$\limsup_{r \to 0} |A_r f - f| \le \epsilon \text{ off a set of measure } \epsilon.$$

Since this holds for all $\epsilon > 0$, we are done. The same proof works for arbitrary A_r. $\square$

Corollary 3.3.9 *Given any $f \in L^p$, $1 \le p \le \infty$, there exists a set S with $m({}^cS) = 0$ such that, for all $x_0 \in S$,*

$$\lim_{r \to 0} \frac{1}{m(B(x_0, r))} \int_{B(x,r_0)} |f(y) - f(x_0)|\, dy = 0.$$

Proof For any $c \in \mathbb{C}$, we have

$$\frac{1}{m(B(x_0, r))} \int_{B(x_0,r)} |f(y) - c|\, dy \to |f(x) - c|$$

almost everywhere. We let c range over rational values and get a countable union of exceptional sets of measure 0. That proves the result. $\square$

Example 3.3.10 Let $B(x, r) \subseteq S$ denote $\{\theta : |\theta - x| < r\}$ for all $r \le \pi$, $x \in S$. Let $B(x, r) = B(x, \pi)$ for $r > \pi$. Recall that

$$P_r(\theta) = \frac{1 - r^2}{1 - 2r\cos\theta + r^2}, \ r < 1 \text{ on the disc.}$$

We have that

$$P_r(\varphi) \le \frac{c\delta}{\delta^2 + \varphi^2} \text{ for } |\varphi| \le \pi \text{ and } \delta = 1 - r.$$

Then

$$|u(r,\theta)| \le \int_{-\pi}^{\pi} P_r(\varphi)|f(\theta - \varphi)|\,d\varphi$$

$$\le C \int_{-\pi}^{\pi} \frac{\delta}{\delta^2 + \varphi^2}|f(\theta - \varphi)|\,d\varphi$$

$$\le \frac{C\delta}{\delta^2} \int_{|\varphi| \le \delta} |f(\theta - \varphi)|\,d\varphi + \sum_k C\delta \int_{2^k\delta \le |\varphi| \le 2^{k+1}\delta} \frac{1}{(2^k\delta)^2}|f(\theta - \varphi)|\,d\varphi$$

$$\le CMf(\theta) + C\sum_k 2^{-k}Mf(\theta)$$

$$\le CMf(\theta).$$

This is an important inequality which holds in much more general settings and for a large class of kernels.

3.4   The Poisson Kernel and Fourier Inversion

Now let f on S be given, and

$$u(r,\theta) = PI(f) = P_r * f = \frac{1}{2\pi} \int_0^{2\pi} P_r(\varphi)f(\theta - \varphi)\,d\varphi$$

with

$$P_r(\varphi) = \frac{1 - r^2}{1 - 2r\cos\varphi + r^2} \; , \; f \in L^p \; , \; 1 \le p \le \infty.$$

We know that $\lim_{r \to 1} u(r,\theta) = f(\theta)$ almost everywhere.

Observation Let $f \in L^p, 1 \le p \le \infty$ and $u = PI(f)$. Then

(1) $\left(\int_0^{2\pi} |u(r,\theta)|^p\,d\theta \right)^{1/p} \le \|f\|_{L^p}$, any $0 < r < 1$.

(2) $\left(\int_0^{2\pi} |u(r,\theta) - f(\theta)|^p\,d\theta \right)^{1/p} \to 0$ as $r \to 1$.

Proof

(1) $\|P_r * f\|_{L^p} \le \|P_r\|_{L^1} \cdot \|f\|_p \le \|f\|_p$.
(2) Let $\omega_p(\varphi) = \|f(\theta - \varphi) - f(\theta)\|_{L^p(\theta, d\theta)}, f \in L^p$. Then

    (a) $|\omega_p(\varphi)| \le 2\|f\|_{L^p}$.
    (b) $\omega_p(\varphi) \to 0$ as $\varphi \to 0$.

    Now

$$u(r, \theta) - f(\theta) = \frac{1}{2\pi} \int_0^{2\pi} P_r(\varphi)[f(\theta - \varphi) - f(\theta)] \, d\varphi$$

so that

$$\|u(r, \theta) - f(\theta)\|_{L^p} \leq \frac{1}{2\pi} \int_0^{2\pi} P_r(\varphi) \omega_p(\varphi) \, d\varphi$$

$$= \int_{|\varphi| \leq \epsilon} + \int_{|\varphi| > \epsilon}$$

$$= T_1 + T_2 \, .$$

As $r \to 1$, T_2 vanishes and T_1 remains small. The proof is completed as in the pointwise case.   □

Remarks on the Poisson Kernel : If $f \in L^1(\mathbb{R})$, then define

$$\widehat{f}(\xi) = \int_{-\infty}^{\infty} e^{2\pi i x \cdot \xi} f(x) \, dx \, .$$

Convolution If $f, g \in L^1$, then

$$(f * g)(x) \equiv \int_{-\infty}^{\infty} f(x - y)g(y) \, dy = \int_{-\infty}^{\infty} f(y)g(x - y) \, dy \, .$$

It holds that

$$\|f * g\|_{L^1} \leq \|f\|_{L^1} \cdot \|g\|_{L^1}$$

and

$$(f * g)^{\widehat{}} = \widehat{f} \cdot \widehat{g} \, .$$

Fourier Inversion Is it true that

$$\int_{-\infty}^{\infty} e^{-2\pi i x \cdot \xi} \widehat{f}(\xi) \, d\xi = f(x) \, ?$$

If $\widehat{f} \in L^1$, then this holds almost everywhere. If $f \in L^1 \cap C$, then it holds everywhere. Recall on the disc that

$$P(r, \theta) = \sum_{-\infty}^{\infty} r^{|n|} e^{in\theta} \, , \quad 0 < r < 1 \, .$$

So we let

$$P_y(x) = \int_{-\infty}^{\infty} e^{-2\pi i x \cdot \xi} e^{-2\pi |\xi| y} \, d\xi = \frac{1}{\pi} \frac{y}{y^2 + x^2}$$

in the upper halfplane $y > 0$.

Note that

$$\int_{-\infty}^{\infty} P_y(x) e^{2\pi i x \cdot \xi} \, dx = e^{-2\pi |\xi| y}$$

by Fourier inversion. If $f \in L^p(\mathbb{R})$ is given, then write

$$u(x, y) = PI(f) = P_y * f(x) = \int_{-\infty}^{\infty} P_y(t) f(x - t) \, dt \ , \ y > 0.$$

So

$$\widehat{u}(\xi, y) = e^{-2\pi |\xi| y} \widehat{f}(\xi)$$

and

$$u(x, y) = \int_{-\infty}^{\infty} e^{-2\pi |\xi| y} e^{-2\pi i x \cdot \xi} \widehat{f}(\xi) d\xi$$

is harmonic on E_2^+. We also have

$$\lim_{y \to 0} u(x, y) \overset{\text{(formally)}}{=} \int_{-\infty}^{\infty} e^{-2\pi i \xi \cdot x} \widehat{f}(\xi) \, d\xi = f(x).$$

Theorem 3.4.1 *Let $f \in L^p(\mathbb{R}^1)$, $1 \leq p \leq \infty$ and set $u = PI(f)$. Then*

(1) $\lim_{y \to 0} u(x, y) = f(x)$ *for almost every x (in particular, the Lebesgue set).*
(2) *For $p < \infty$, $u(x, y) \to f(x)$ in L^p.*
(3) $|u(x, y)| \leq CMf(x)$ *where*

$$Mf(x) = \sup_\delta \frac{1}{2\delta} \int_{-\delta}^{\delta} |f(x - y)| \, dy.$$

n.b. On n-dimensional domains, e.g., the n-torus, there are no longer explicit formulas for the Poisson kernel. We will want to deal with this problem directly, so let us consider how the Poisson kernel arises in $\mathbb{R}^1$ (or S).

(A) Fourier series: If $f \sim \sum a_n e^{in\theta}$, then $u(r, \theta) = \sum a_n r^{|n|} e^{in\theta}$.
(B) Green's theorem: Let $\mathcal{D} \subseteq \mathbb{R}^2$ be bounded and have "smooth" boundary. We seek $G(z, \varsigma)$, the Green's function, on $\mathcal{D} \times \mathcal{D} \setminus$ diagonal. Now G is uniquely determined by

(1) For each z, G is harmonic in ζ and "regular" up to $\partial\mathcal{D}$.

(2) $G(z,\zeta) = (-1/(2\pi))\log|z - \zeta| + \text{(regular when } \zeta = z\text{)}$.

(3) $G(z,\zeta)\Big|_{\zeta\in\partial\mathcal{D}} = 0$.

One can construct such a G using the Perron solution of the Dirichlet problem. One can show, using Green's theorem, that G is symmetric in z, ζ. One then defines $P(z,\zeta) = (\partial/\partial\eta_\zeta)G(z,\zeta)$ at $\zeta \in \partial\mathcal{D}$, where η_ζ is the unit outward normal vector at ζ.

Now Green's formula says that if $u, v \in C^2(\mathcal{D}) \cap C(\overline{\mathcal{D}})$, then

$$\int_{\mathcal{D}} u\Delta v - v\Delta u\, dA = \int_{\partial\mathcal{D}} u\frac{\partial v}{\partial\eta_\zeta} - v\frac{\partial u}{\partial\eta_\zeta}\,.$$

If $u \in h(\mathcal{D}) \cap C(\overline{\mathcal{D}})$, then letting $v(\zeta) = G(z,\zeta)$ yields

$$\int_{\partial\mathcal{D}} u(\zeta)P(z,\zeta)\,d\zeta = u(z),$$

that is, P has the reproducing property for functions in $u \in h(\mathcal{D}) \cap C(\overline{\mathcal{D}})$.

We remark that, on the unit disc, it is not difficult to calculate that

$$G(z,\zeta) = -\frac{1}{2\pi}\left(\log|z - \zeta| - \log|z - \overline{\zeta}^{-1}|\right).$$

(C) The Cauchy kernel: Let $C(z,\zeta) = $ the Cauchy-Szegő kernel for the unit disc,

$$C(z,\zeta) = \frac{1}{2\pi}\frac{1}{1 - z\cdot\overline{\zeta}}\,,\ |z| < 1\,,\ |\zeta| < 1\,.$$

If $F \in H(D) \cap C(\overline{D})$, then

$$F(z) = \frac{1}{2\pi}\oint\frac{F(\zeta)\,d\zeta}{\zeta - z} = \int_{|\zeta|=1} F(\zeta)C(z,\zeta)\,ds_\zeta\,.$$

Here ds_ζ is arc-length measure on the boundary of the domain.

Define

$$\frac{1}{2\pi}P(z,\zeta) = \frac{|C(z,\zeta)|^2}{C(z,z)}\,.$$

We claim that if $F \in H(D) \cap C(\overline{D})$, then

$$F(z) = \int_{|\zeta|=1} P(z,\zeta)F(\zeta)\,d\sigma(\zeta)\,. \tag{$*$}$$

[Here $d\sigma$ is the induced Euclidean measure on ∂D.]

To prove (∗), let $z_0 \in D$ be fixed. Let

$$g(z) = F(z) \cdot \overline{C(z_0, z)}.$$

Then $g \in H(D) \cap C(\overline{D})$, so we apply the reproducing property of C to g at z_0. Thus

$$g(z) = F(z) \cdot \overline{C(z_0, z)} = \int_{|\zeta|=1} C(z, \zeta) F(\zeta) \overline{C(z_0, \zeta)} \, d\sigma(\zeta).$$

Let $z = z_0$ so that

$$g(z_0) = F(z_0) \cdot \overline{C(z_0, z_0)} = \int_{|\zeta|=1} |C(z_0, \zeta)|^2 F(\zeta) \, d\sigma(\zeta)$$

or

$$F(z_0) = \int_{|\zeta|=1} \frac{|C(z_0, \zeta)|^2}{C(z_0, z_0)} F(\zeta) \, d\sigma(\zeta).$$

Since (∗) applies to functions in $H(D) \cap C(\overline{D})$, it applies surely to z^n. But since $P(z, \zeta) \geq 0$, it thus applies to $\operatorname{Re} z^n$ and $\operatorname{Im} z^n$, hence to all harmonic functions. [n.b. An argument in the reverse direction would not work since the Hilbert transform does not map continuous functions to continuous functions.] Equation (∗) provides a useful version of P for several complex variables. It differs materially from **(B)** for $n > 1$.

(D) Via the conformal group.

Consider the group of linear fractional transformations on D. Any element of this group can be written as

$$\varphi_{ab} : z \mapsto e^{ia}\left(\frac{z - b}{1 - \overline{b}z}\right) , \quad a \in \mathbb{R} , \quad |b| < 1.$$

The $\{\varphi_{a0}\}$ form a subgroup: the rotations. The group of linear fractional transformations, which we denote by $\mathcal{G}$, is transitive on D.

Proof φ_{0x} maps x to 0.   □

Harmonic functions are invariant under linear fractional transformations. For let $\alpha \in \mathcal{G}$ and suppose that $u \in h(D)$. Let

$$P(z, \zeta) = \frac{1}{2\pi} P_r(\theta - \varphi) , \quad z = re^{i\theta} , \quad \zeta = \epsilon^{i\varphi}.$$

Now

$$\partial\bar{\partial}u \circ \alpha(z) = \partial\left[\frac{\partial u}{\partial\bar{z}}\frac{\partial\bar{\alpha}}{\partial z}\right]$$

$$= \left[\frac{\partial^2 u}{\partial z\partial\bar{z}}\cdot\frac{\partial\alpha}{\partial z}\cdot\frac{\partial\bar{\alpha}}{\partial\bar{z}}\right]$$

$$= 0.$$

So $u \circ \alpha \in h(D)$. Now if also $u \in C(\bar{D})$, then

$$u(z) = \int_{|\zeta|=1} P(z,\zeta)u(\zeta)\,d\sigma_\zeta$$

and

$$u(0) = \frac{1}{2\pi}\int_0^{2\pi} u(e^{i\theta})\,d\theta.$$

Since $u \circ \alpha \in C(\bar{D})$, we have also that

$$u(\alpha(0)) = \frac{1}{2\pi}\int_0^{2\pi} u(\alpha(e^{i\theta}))\,d\theta$$

for any $\alpha \in \mathcal{G}$. Now suppose that α is a member of the coset of linear fractional transformations which take 0 to $z, z \in D$ fixed. Thus

$$u(z) = u(\alpha(0)) = \frac{1}{2\pi}\int_0^{2\pi} u(\alpha(e^{i\theta}))\,d\theta.$$

Let $e^{i\varphi} = \alpha(e^{i\theta})$, or $e^{i\theta} = \alpha^{-1}(e^{i\varphi})$. Then

$$\left|\frac{d\theta}{d\varphi}\right| = \mathcal{J}_{\alpha^{-1}}(e^{i\varphi})$$

and thus

$$u(z) = \int_0^{2\pi} u(e^{i\varphi})\mathcal{J}_{\alpha^{-1}}(e^{i\varphi})\frac{d\varphi}{2\pi}.$$

Here $\mathcal{J}$ is of course the Jacobian determinant of the mapping.
    By the uniqueness of the Poisson kernel,

$$P(z,\zeta) = \frac{1}{2\pi}\mathcal{J}_{\alpha^{-1}}(\zeta),$$

Fig. 3.6 Comparison of the
disc and the upper halfplane

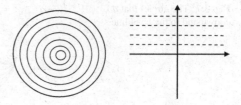

where $\alpha \in \mathcal{G}$ is such that $\alpha : 0 \mapsto z$. This is the geometric interpretation of
P. It is the basic idea underlying the notion of harmonic functions and Poisson
integrals for symmetric spaces.

(E) One can also obtain the Poisson kernel for D from the Poisson kernel for the
unbounded realization of D (the upper halfplane). We are after

$$P(z, \zeta) = \frac{1}{2\pi} \frac{1 - |z|^2}{|z - \zeta|^2} \ , \ |z| \le 1 \, , \ |\zeta| = 1 \, ,$$

to be obtained from

$$P(z, \zeta) = \frac{1}{\pi} \frac{y}{y^2 + |x - \zeta|^2} \ , z = x + iy \, , \ y > 0 \, , \ \zeta \in \mathbb{R} \, .$$

Recall the Cayley transform

$$\gamma : z \mapsto \frac{1}{i} \cdot \frac{z - 1}{z + 1}$$

with

$$\gamma^{-1} : w \mapsto \frac{i - w}{i + w} \, .$$

Note that

$$\gamma(1) = 0 \ , \ \ \gamma(-1) = \infty \ , \ \ \gamma(0) = i \, .$$

Of course γ preserves harmonic functions so one would expect that it would
pull the Poisson kernel from the halfplane back to the disc. There are problems
with this approach. Refer to the figure as we discuss the matter (Fig. 3.6).

The circles on the left do *not* go to the level lines on the right under γ, so
the H^p spaces do not correspond canonically. Instead, these circles in the next
figure correspond to these lines (Fig. 3.7).

In spite of this obstruction, the two H^p spaces are indeed the same. And we
have

$$P_D(z, \zeta) = P_{\mathbb{R}_+^2}(\gamma(z), \gamma(\zeta))\mathcal{J}_\gamma(\zeta) \, .$$

Fig. 3.7 The circles that *do* correspond to the level lines

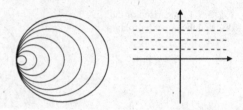

This approach is also significant in the theory of symmetric spaces.

(F) Periodization (important in passing from $\mathbb{R}^n$ to $\mathbb{T}^n$).

Motivation Let $u(r, \theta) \in h(D) \cap L^2(S)$, with

$$u(r, \theta) = \sum a_n |r|^n e^{in\theta} .$$

Let $r = e^{-y}, 0 < y < \infty$ and $\theta = x$. So

$$u(r, \theta) = \sum a_n e^{-|n|y} e^{inx} \in h(\mathbb{R}^2_+)$$

and is periodic in x.

Suppose that a function f on $\mathbb{R}^1$ is given. What is its periodic analogue? It is

$$\sum_{-\infty}^{\infty} f(x + 2n\pi) \equiv f_p(x) .$$

So one might expect that

$$P_r(\theta) = \sum_{-\infty}^{\infty} P_{\mathbb{R}^2_+} (t, \theta/(2\pi) + n) \;, r = e^{-2\pi t} .$$

Chapter 4
Hardy Spaces on D

4.1   The Role of the Hilbert Transform

We begin with $p = 2$ since this reflects the formal structure of the situation, but the mechanics are easier.

Recall that

$$H^2(D) = \{F \in H(D) : \sup_{1 > r > 0} \int_0^{2\pi} |F(re^{i\theta})|^2 \, d\theta \equiv \|F\|_{H^2}^2 < \infty\}.$$

Write $F = u + iv$, $u, v \in h(D)$ and real-valued. We have

$$F(z) = \sum_{n=0}^{\infty} c_n z^n = \sum_{n=0}^{\infty} c_n r^n e^{in\theta}.$$

Now $F \in H^2$ if and only if $\sum |c_n|^2 < \infty$. In fact

$$\frac{1}{2\pi} \|F\|_{H^2}^2 = \sum_{n=0}^{\infty} |c_n|^2.$$

Of course, $u(r, \theta) = \mathrm{Re}\, F$ is harmonic, and

$$\sup_{r < 1} \int_0^{2\pi} |u(r, \theta)|^2 \, d\theta < \infty,$$

so that

$$u = PIf \quad \text{with } f \in L^2(S).$$

© The Author(s), under exclusive license to Springer Nature Switzerland AG 2023
S. G. Krantz, *The E. M. Stein Lectures on Hardy Spaces*, Lecture Notes in
Mathematics 2326, https://doi.org/10.1007/978-3-031-21952-8_4

Thus

$$u(r, \theta) = \sum_{n=-\infty}^{\infty} a_n r^{|n|} e^{in\theta}.$$

What is the relationship between a_n and c_n? In fact,

$$a_n = \frac{c_n}{2} \quad \text{if } n > 0$$

and

$$a_n = \frac{\overline{c_{-n}}}{2} \quad \text{if } n < 0$$

and

$$z_0 = \frac{c_0 + \overline{c_0}}{2}.$$

Write

$$v(r, \theta) = \sum_{-\infty}^{\infty} b_n r^{|n|} e^{in\theta}.$$

Then

$$b_n = \frac{\operatorname{sgn} n}{i} a_n \quad \text{if } n \neq 0$$

and

$$b_0 = \frac{c_0 - \overline{c_0}}{2}.$$

The Conjugate Function Suppose that $f \in L^1(-\pi, \pi)$ is complex-valued, $f \sim \sum a_n e^{in\theta}$ (n.b. f is real-valued if and only if $\overline{a_n} = a_{-n}$ for every n). Write

$$\widetilde{f} \sim \sum_{n \neq 0} \frac{a_n \operatorname{sgn} n}{i} e^{in\theta}.$$

Consider the operator $f \mapsto \widetilde{f}$, which is the *Hilbert transform*. This operator is clearly well defined on L^2 and

$$\|\widetilde{f}\|_2^2 + 2\pi a_0^2 = \|f\|_2^2.$$

Note that

$$\widetilde{\widetilde{f}} = -f .$$

Proposition 4.1.1 *The function* $F \in H^2$ *if and only if there exists a real-valued* $f \in L^2(S)$ *and a real constant* b_0 *such that*

$$F = PIf + iPI\widetilde{f} + ib_0$$

and

$$\|F\|_{H^2}^2 = \|f\|_2^2 + \|\widetilde{f}\|_2^2 + 2\pi b_0^2 .$$

Recall the Poisson kernel $P(r, \theta)$:

$$\sum r^{|n|} e^{in\theta} = \frac{1 - r^2}{1 - 2r\cos\theta + r^2} .$$

Define

$$Q(r, \theta) = \sum_{n \neq 0} r^{|n|} \frac{\operatorname{sgn} n}{i} e^{in\theta} = \frac{2r\sin\theta}{1 - 2r\cos\theta + r^2} .$$

Proposition 4.1.2 *If* $f \in L^2(S)$ *(so* $\widetilde{f} \in L^2(S)$*), then* $P_r * \widetilde{f} = PI(\widetilde{f}) = Q_r * f.$

Proof Check the Fourier coefficients of both sides.   □

By Proposition 4.1.2, if $f \in L^2$, then $\lim_{r \to 1} Q_r * f = \widetilde{f}$ almost everywhere and in L^2 norm.

Now

$$Q_r * f(\theta) = \frac{1}{2\pi} \int_{-\pi}^{\pi} f(\theta - \varphi) \frac{2r\sin\varphi}{1 - 2r\cos\varphi + r^2} .$$

Formally at least, as $r \to 1$, this last expression tends to

$$\frac{1}{2\pi} \int_{-\pi}^{\pi} f(\theta - \varphi) \frac{2\sin\varphi}{2 - 2\cos\varphi} d\varphi = \int_{-\pi}^{\pi} \frac{1}{2\pi} \frac{\sin\varphi}{1 - \cos\varphi} f(\theta - \varphi) d\varphi$$

$$= \int_{-\pi}^{\pi} \frac{1}{2\pi} \cdot \frac{2\sin(\varphi/2)\cos(\varphi/2)}{2(\sin(\varphi/2))^2} f(\theta - \varphi) d\varphi$$

$$= \frac{1}{2\pi} \int_{-\pi}^{\pi} \cot(\varphi/2) f(\theta - \varphi) d\varphi .$$

As usual, let $\delta = 1 - r$. Define

$$T_\delta f(\theta) = \frac{1}{2\pi} \int_{\pi \geq |\varphi| > \delta} f(\theta - \varphi) \cot \frac{\varphi}{2} \, d\varphi \, .$$

Proposition 4.1.3 *We have:*

(a) *If $f \in L^2$, then $\lim_{\delta \to 0} T_\delta f = \tilde{f}$ almost everywhere and in L^2.*

(b) *If $f \in L^p$, $1 \leq p \leq \infty$, then $Q_r * f - T_\delta f \to 0$ almost everywhere. When $p < \infty$, then the convergence takes place in L^p norm as well.*

Proof It suffices to prove (b). Define

$$K_r(\varphi) = \begin{cases} Q_r(\varphi) & \text{if} \quad |\varphi| \leq 1 - r \\ Q_r(\varphi) - Q_1(\varphi) & \text{if} \quad |\varphi| > 1 - r \, . \end{cases}$$

Then

$$Q_r * f - T_\delta f = K_r * f = \int_{-\pi}^{\pi} \frac{1}{2\pi} K_r(\varphi) f(\theta - \varphi) \, d\varphi \, .$$

Trivially, $\int_{-\pi}^{\pi} K_r(\varphi) \, d\varphi = 0$ for all r, so this last

$$= \int K_r(\varphi) [f(\theta - \varphi) - f(\theta)] \, d\varphi \, .$$

Once one sees that $|K_r(\varphi)| \leq c\delta/(\delta^2 + \varphi^2)$ with $\delta = 1 - r$, then one can complete the proof as usual. But, for $|\varphi| > 1 - r$,

$$\begin{aligned}
|K_r(\varphi)| &= |Q_r(\varphi) - Q_1(\varphi)| \\
&= \left| \frac{2r \sin \varphi}{1 - 2r \cos \varphi + r^2} - \frac{2 \sin \varphi}{2 - 2\cos \varphi} \right| \\
&= \left| \frac{2r \sin \varphi}{(1-r)^2 + 2r(1 - \cos \varphi)} - \frac{\sin \varphi}{1 - \cos \varphi} \right| \\
&= \left| \frac{(1-r)^2 \sin \varphi}{(1 - \cos \varphi)[(1-r)^2 + 2r(1 - \cos \varphi)]} \right| \\
&\leq \left| \frac{(1-r)^2}{(1-r)[(1-r)^2 + 2r(1 - \cos \varphi)]} \right| \\
&\leq \left| \frac{1-r}{(1-r)^2 + 2r(1 - \cos \varphi)} \right| \\
&= P_r(\varphi) \\
&\leq \frac{C\delta}{\delta^2 + \varphi^2} \, .
\end{aligned}$$

When $|\varphi| \leq 1 - r$, we have

$$
\begin{aligned}
|K_r(\varphi)| &= |Q_r(\varphi)| \\
&= \left| \frac{2r \sin \varphi}{1 - 2r \cos \varphi + r^2} \right| \\
&\leq \left| \frac{2(1 - r)}{1 - 2r \cos \varphi + r^2} \right| \\
&\leq 2P_r(\varphi) \\
&\leq \frac{2C\delta}{\delta^2 + \varphi^2} \cdot
\end{aligned}
$$

That completes the proof.   □

Now, at least, we know it is consistent to define

$$
\tilde{f} \sim \sum_{n \neq 0} a_n \frac{\operatorname{sgn} n}{i} e^{in\theta}
$$

or

$$
\tilde{f} = \lim_{\delta \to 0} \frac{1}{2\pi} \int_{|\varphi| \geq \delta} f(\theta - \varphi) \cot \frac{\varphi}{2} \, d\varphi
$$

whenever either of them makes sense and $f \sim \sum a_m e^{in\theta}$.

We now have:

Theorem 4.1.4 (M. Riesz) *For $1 < p < \infty$:*

(1) *If $f \in L^p$, then $\lim_{\delta \to 0} T_\delta f = f$ almost everywhere and in L^p norm.*
(2) *We have $\|Q_r * f\|_{L^p} \leq A_p \|f\|_{L^p}$ with A_p independent of $r < 1$ and also independent of f.*
(3) *The function $F \in H^p$ if and only if there exists a real-valued $f \in L^p$ and a real constant b_0 such that*

$$
F = PIf + iQ_r * f + ib_0
$$

and

$$
\|F\|_{H^p} \approx \|f\|_{L^p} + |b_0| .
$$

We defer the proof for a while so that we can make some remarks and pose some open problems.

Remark The result of Riesz is not valid when $p = \infty$ or $p = 1$. No similar result holds in those cases. For example, when $p = \infty$, let

$$F(z) = \log\left(\frac{1}{1-z}\right) = \sum_{n=1}^{\infty} \frac{z^n}{n}.$$

The only singularity of F is at $z = 1$.

Now

$$\text{Im } F \sim \sum_{n=1}^{\infty} \frac{\sin n\theta}{n} r^n = \arctan\left(\frac{r \sin\theta}{1 - r\cos\theta}\right) \to f(\theta)$$

and

$$\text{Re } F \sim \sum_{n=1}^{\infty} \frac{\cos n\theta}{n} r^n = \frac{1}{2}\log\left(\frac{1}{1 - 2r\cos\theta + r^2}\right) \to \tilde{f}(\theta).$$

So

$$f(\theta) = \frac{1}{2}|\pi - \theta| \ , \ 0 < \theta < 2\pi,$$

and

$$\tilde{f}(\theta) = -\log(2|\sin(\theta/2)|) \ , \ |\theta| < \pi .$$

Hence, clearly $f \in L^\infty$, $\tilde{f} \notin L^\infty$. By duality, this gives a counterexample for L^1 as well.

We digress to state some unsolved problems and then return to give two proofs of Riesz's theorem.

PROBLEM 1 (SPECIAL CASE) This problem is in $\mathbb{R}^n$ with $n = 2$. Let $E \subseteq S$ be measurable, $m(E) > 0$. Set

$$D(r, E) = \{z : |z| \le r, \arg z \in E\}.$$

Observe that $m(D(r, E)) = m(E)r^2/2$. Define

$$M_E(f) = \sup_{r>0} \frac{1}{m(D(r, E))} \int_{D(r,E)} f(x - y)\,dy .$$

We claim that M_E satisfies L^p inequalities:

(1) $\|M_E f\|_{L^p} \le A(p, E) \cdot \|f\|_{L^p} \ , \ 1 < p \le \infty.$
(2) $m\{x : M_E f(x) > \alpha\} \le \frac{A(1, E)}{\alpha}\|f\|_{L^1}.$

These are clear since

$$M_E f(x) \leq \frac{1}{|E|} M f(x).$$

It can also be shown, but it is nontrivial, that

$$\|M_E f\|_{L^p} \leq A_p \|f\|_{L^p}$$

with A_p independent of E. This uses the method of rotations.

Question What is the analogue for $p = 1$? That is, is $A(1, E) \leq A(1)$? A weaker question is, does there exist an α, $0 \leq \alpha < 1$, such that

$$A(1, E) \leq \frac{A}{[m(E)]^\alpha} ?$$

As references, see [CAZ1, CAZ2]. Also E. M. Stein [STE3].

Proof of Riesz's Theorem It is not difficult to check that the three statements of the theorem are logically equivalent. We give two proofs of **(2)**.

First Proof (due to Riesz himself): This works easiest if p is even: $p = 2k, k = 1, 2, \ldots$. Easiest of all is $p = 2$. Assume without loss of generality that $f \geq 0$. Write

$$F(z) = PI(f) + i QI(f).$$

(We will forget about the b_0 term.) So $F \in H(D)$. Thus $F^2 \in H(D)$ and

$$\frac{1}{2\pi} \int_0^{2\pi} F^2(re^{i\theta}) \, d\theta = F^2(0) = a_0^2.$$

Equating real parts, we get

$$\frac{1}{2\pi} \int_0^{2\pi} (u^2(r, \theta) - v^2(r, \theta)) \, d\theta = a_0^2$$

or

$$\frac{1}{2\pi} \int_0^{2\pi} v^2(r, \theta) \, d\theta = \frac{1}{2\pi} \int_0^{2\pi} u^2(r, \theta) \, d\theta - a_0^2.$$

This gives the L^2 identity. For $p = 4$, we have, assuming as we may have that $\|f\|_{L^4} = 1$, that

$$\frac{1}{2\pi} \int_0^{2\pi} F^4(re^{i\theta}) \, d\theta = a_0^4.$$

But

$$\operatorname{Re} F^4 = u^4 - 6u^2 v^2 + v^4$$

so

$$\frac{1}{2\pi} \left(\int_0^{2\pi} u^4 + \int_0^{2\pi} v^4 - 6 \int_0^{2\pi} u^2 v^2 \right) = a_0^4 .$$

(If the third term on the left did not appear, then we would be finished—just as in the L^2 case.)

Now define

$$\mathcal{J}^2(r) = \frac{1}{2\pi} \int_0^{2\pi} u^2(r, \theta) \cdot v^2(r, \theta) \, d\theta .$$

This is the term we wish to estimate. So

$$\mathcal{J}^2(r) \le a_0^4 + 6 \cdot \frac{1}{2\pi} \int_0^{2\pi} u^2 v^2 \, d\theta$$

$$\le a_0^4 + C \int u^4 \, d\theta^{1/2} \cdot \int v^4 \, d\theta^{1/2}$$

$$\le a_0^4 + C \|f\|_4^2 \cdot \mathcal{J}(r) .$$

So we have

$$\mathcal{J}^2(r) \le c_1 \mathcal{J}(r) + c_2 .$$

In conclusion, $\mathcal{J}(r) \ge 1$ implies that

$$\mathcal{J}(r) \le c_1 + \frac{c_2}{\mathcal{J}(r)} \le c_1 + c_2 .$$

Hence,

$$\mathcal{J}(r) \le 1 + c_1 + c_2 .$$

End of proof.

One can duplicate this procedure for any even $p \ge 2$ and interpolate to obtain Riesz's estimate for all $p \ge 2$, $p \ne \infty$. By duality, one can obtain the result for $1 < p < 2$.   □

Now we give a proof using subharmonic functions. This approach will be important in the sequel.

Definition 4.1.5 Suppose that $\mathcal{R} \subseteq \mathbb{R}^n$ is open, $f \in C(\mathcal{R})$, and f real-valued. We say that f is *subharmonic* on $\mathcal{R}$ if, whenever $\mathcal{R}_1 \subseteq \mathcal{R}$ is open and relatively compact, and $u \in h(\mathcal{R}_1) \cap C(\overline{\mathcal{R}_1})$, $f \leq u$ on $\partial \mathcal{R}_1$, then $f \leq u$ on $\mathcal{R}_1$.

Trivial Example When $n = 1$, the harmonic functions are the linear ones and the subharmonic functions are the convex ones. (Note that convexity is characterized by $d^2 f / dx^2 \geq 0$—this will come up again later.)

Lemma 4.1.6 *Let $f \in C^2(\mathcal{R})$. Then f is subharmonic on $\mathcal{R}$ if and only if $\Delta f \geq 0$.*

This lemma follows from a "maximum principle": If f is given on $\overline{\mathcal{R}_1}$, $\overline{\mathcal{R}_1}$ compact, $f \in C(\overline{\mathcal{R}_1}) \cap C^2(\mathcal{R}_1)$, $\Delta f \geq 0$ on $\mathcal{R}_1$, and $f \leq 0$ on $\partial \mathcal{R}_1$, then $f \leq 0$ on $\mathcal{R}_1$ (f here is assumed to be real-valued).

Proof of the Main Result Assume first that $\Delta f > 0$ on $\mathcal{R}_1$, and suppose that $f(z_0) > 0$, some $z_0 \in \mathcal{R}_1$. We seek a contradiction.

Choose $x_0 \in \mathcal{R}_1$ such that $f(x_0)$ is maximal. Now $\Delta f(x_0) > 0$ so that

$$\frac{\partial^2 f}{\partial x_j^2}(x_0) > 0,$$

some j. Let

$$h_j = (0, 0, \ldots, 0, \delta, 0, \ldots, 0),$$

where the δ lives in the jth component. So

$$f(x_0 + h_\delta) = f(x_0) + \frac{\delta^2}{2} \cdot \frac{\partial^2 f}{\partial x_j^2}(x_0) + o(|h_j|^2).$$

For δ small, this gives $f(x_0 + h_\delta) > f(x_0 + h)$, a contradiction.

If only $\Delta f \geq 0$, then consider $f + \epsilon |x|^2 - \delta$, with ϵ and δ small and to be chosen. Fix $\epsilon > 0$, and choose δ so that $f + \epsilon |x|^2 - \delta \leq 0$ on $\partial \mathcal{R}_1$. Then $\Delta(f + \epsilon |x|^2 - \delta) > 0$ and, by the first part of the proof, $f + \epsilon |x|^2 - \delta \leq 0$ on $\mathcal{R}_1$. Let ϵ, δ tend to 0 to get $f \leq 0$ on $\mathcal{R}_1$.

Now Let Us Prove the Lemma If $u \in h(\mathcal{R}_1)$, $u \geq f$ on $\partial \mathcal{R}_1$, look at $f - u$. Now $\Delta(f - u) \geq 0$. So, by the above, $f - u \leq 0$ in $\mathcal{R}_1$, that is, $f \leq u$.

For the converse, suppose that $f \in C^2$ and f satisfies harmonic majorization. If $x \in \mathcal{R}_1$, u harmonic near x, then

$$u(x) = \frac{1}{2\pi} \int_{|y'|=r} u(x + ry') \, dy'$$

for r small. Let

$$\mathcal{J}(r) = \frac{1}{2\pi} \int_{|y'|=r} u(x + ry') \, dy'$$

for r small. One has

$$\mathcal{J}(0) \le \mathcal{J}(r)$$

by the mean value property. Thus

$$0 \le \frac{d^2}{dr^2} \mathcal{J}(r) \bigg|_{r=0} = C \triangle f(x)$$

for $C \ge 0$. Thus $\triangle f(x) \ge 0$ as desired.   □

Observation Suppose that $u \in h(\mathcal{R})$, $u \ge 0$. Then u^p, $p \ge 1$, is subharmonic. For

$$|u(x)|^p = \left| \frac{1}{2\pi} \int_{|y'|=1} u(x + ry') \, dy' \right|^p \le \frac{1}{2\pi} \int_{|y'|=1} u^p(x + ry') \, dy'$$

by Jensen's inequality.

Now we compute $\triangle(|u|^p)$. This will of course be ≥ 0, but it is useful to know it explicitly.

Lemma 4.1.7 *Suppose that $u \in h(\mathcal{R})$, $u > 0$. Then*

$$\triangle(u^p) = p(p-1)u^{p-2}|\nabla u|^2 .$$

Note There cannot be any second order terms on the right because the expression must be rotationally invariant. And the only second order differential operator that is rotationally invariant is $\triangle$; and in our case, $\triangle u = 0$. Note that ∇ is a first order, rotationally invariant differential operator.

Proof of the Lemma Now

$$\frac{\partial}{\partial x_j} u^p = p u^{p-1} u_j ,$$

where the subscript j indicates a derivative in the jth variable. Therefore,

$$\frac{\partial^2}{\partial x_j^2} u^p = p(p-1)u^{p-2}(u_j)^2 + p u^{p-1} u_{jj} .$$

Hence

$$\sum_j \frac{\partial^2}{\partial x_j^2}(u^p) = p(p-1)u^{p-2}|\nabla u|^2 .$$

Observe that, for $p < 1$, the identity still holds; but the expression on the right is no longer positive.

The corresponding identity for $F \in H(\mathbb{C}^1)$ is this: If $F \in H(\mathbb{C}^1)$, F does not vanish, then

$$\Delta(|F(z)|^p) = p^2|F|^{p-2} \cdot |F'|^2 .$$

This is the Hardy–Stein identity. (But *not* E. M. Stein. In fact, this was some obscure Stein who was never heard from again.) Now, for holomorphic functions, the positivity is good for all positive p. So $|F|^p$ is subharmonic for all $p > 0$. This will be true even if F vanishes, but the proof then is more delicate.

Proof of the Hardy–Stein Result Write

$$|F(z)|^p = F(z)^{p/2} \cdot \overline{F(z)}^{p/2} .$$

Now $\Delta = 4(\partial/\partial z)(\partial/\partial \overline{z})$. Hence,

$$\frac{1}{4}\Delta|F(z)|^p = \overline{\partial}_z \partial_z (F^{p/2} \cdot \overline{F}^{p/2})$$

$$= \overline{\partial}_z \left(\overline{F}^{p/2} \cdot (p/2) \cdot F^{p/2-1} \cdot F' \right)$$

$$= \frac{p}{2} \cdot \overline{F}^{p/2-1} \cdot \overline{F}' \cdot (p/2) \cdot F^{p/2-1} \cdot F'$$

$$= \frac{p^2}{4} \cdot |F|^{p-2} \cdot |F'|^2 ,$$

as desired.   □

We now have enough machinery to give a second proof of the M. Riesz theorem.

Proof Suppose that $f \in L^p(0, 2\pi)$, $1 < p < \infty$. We want $\|Q_f * f\|_{L^p} \leq A_p\|f\|_{L^p}$, $0 < r < 1$. First restrict attention to $1 < p \leq 2$, and suppose that $f \geq 0$. Write $u = PI(f) = u(re^{i\theta}) = P_r * f$, $v = Q_r * f$. Let $F(z) = u(z) + iv(z)$. Since $f \geq 0$, so also is $u > 0$ on the disc. So $F \neq 0$ on the disc.

We consider $|F|^p$ and $|u|^p = u^p$. Now, by the Cauchy–Riemann equations, $|\nabla u|^2 = |F'|^2$, and of course, $|u| \leq |F|$. So, by the Hardy–Stein lemma,

$$\Delta(|F|^p) = p^2|F|^{p-2}|F'|^2$$

$$\leq p^2|u|^{p-2}|\nabla u|^2 \quad \text{since } p \leq 2$$

$$\leq \frac{p}{p-1}\Delta(u^p).$$

By Green's theorem for the disc, if $G \in C^2(D)$, then

$$\int_{|z| \leq r} \Delta G \, dx dy = \frac{d}{dr}\mathcal{J}(r),$$

where

$$\mathcal{J}(r) \equiv \int_0^{2\pi} G(re^{i\theta}) \, d\theta.$$

Let

$$G = \left(\frac{p}{p-1}\right) u^p - |F|^p.$$

So

$$\frac{d}{dr}\mathcal{J}_r = \int_{|z| \leq r} \Delta G \geq 0,$$

and this says

$$\frac{p}{p-1}\frac{d}{dr}\int_0^{2\pi} |u(re^{i\theta})|^p \, d\theta \geq \frac{d}{dr}\int_0^{2\pi} |F(re^{i\theta})|^p \, d\theta . \qquad (*)$$

Now

$$u(re^{i\theta}) = \sum_n a_n r^{|n|} e^{in\theta}$$

and

$$v(re^{i\theta}) = \sum_n \frac{a_n \operatorname{sgn} n}{i} r^{|n|} e^{in\theta}.$$

Furthermore,

$$u(0) = \frac{1}{2\pi}\int_0^{2\pi} f(\theta) \, d\theta \quad \text{and} \quad v(0) = 0.$$

Thus

$$\frac{p}{p-1}\int_0^{2\pi} |u(re^{i\theta}|^p \, d\theta \geq \int_0^{2\pi} |F(re^{i\theta})|^p \, d\theta \text{ at } r = 0.$$

This, coupled with (∗), says that

$$\frac{p}{p-1}\int_0^{2\pi} |u(re^{i\theta})|^p \, d\theta \geq \int_0^{2\pi} |F(re^{i\theta})|^p \, d\theta \text{ for all } r \geq 0.$$

So

$$\frac{p}{p-1}\|f\|_p^p \geq \int_0^{2\pi} |F(re^{i\theta})|^p \, d\theta \geq \|v\|_p^p,$$

provided that $1 < p \leq 2$ and $f \geq 0$.

The case of f not positive follows by writing

$$f = f_1 - f_2 + if_3 - if_4$$

with all $f_j \geq 0$. The case $p > 2$ follows by duality.   □

Lemma 4.1.8 *If $f \in L^p(S)$, $g \in L^q(S)$, $1/p + 1/q = 1$, $1 < p \leq 2$, then*

$$\frac{1}{2\pi}\int_{-\pi}^{\pi} Q_r(f)(\theta) \cdot g(\theta) \, d\theta = -\frac{1}{2\pi}\int_{-\pi}^{\pi} f(\theta) \cdot Q_r(g)(\theta) \, d\theta .$$

Proof If $f \sim \sum a_n e^{in\theta}$ and $g \sim \sum b_n e^{in\theta}$, then of course

$$Q_r f \sim \sum a_n r^{|n|} \frac{\operatorname{sgn} n}{i} e^{in\theta} \text{ for } 0 < r < 1.$$

Hence,

$$\frac{1}{2\pi}\int_{-\pi}^{\pi} Q_r(f)\overline{g} \, d\theta = \sum a_n \overline{b}_n r^{|n|} \frac{\operatorname{sgn} n}{i} = -\frac{1}{2\pi}\int_{-\pi}^{\pi} f \cdot \overline{Q_r(g)} \, d\theta .$$

(Note that $q \geq 2$ so that $g \in L^q \subseteq L^2$; hence, $Q_r g$ makes sense.)   □

Now we have shown that

$$\|Q_r * f\|_{L^p} \leq A_p \|f\|_{L^p}$$

with $A_p \leq 2(p/(p-1))^{1/p}$. Note that, as $p \to 1$, the constant blows up. A year ago, someone found the best constant.

If A_p were bounded as $p \to 1$, the inequality would persist for $p = 1$. But that is not so. It is the case that $A_p \sim 1/(p-1)$ as $p \to 1$.

Now suppose that $2 \leq p < \infty$ so that $1 < q \leq 2$. Then

$$\|Q_r f\|_{L^p} = \sup_{\substack{g \in L^q \\ \|g\|_q \le 1}} \left| \int Q_r(f) \cdot g \, d\theta \right|$$

$$= \sup_g \left| \int f \cdot Q_r(g) \, d\theta \right|$$

$$\le \sup_g \|f\|_p \cdot \|Q_r(g)\|_q$$

$$\le \sup_g \|f\|_p \cdot A_q \cdot \|g\|_q$$

$$\le A_q \cdot \|f\|_p$$

as desired.

Having given two proofs of statement **(2)** of Riesz's theorem, we now indicate why **(1)** and **(3)** follow.

Recall that

$$T_\delta f = \frac{1}{2\pi} \int_{\pi \ge |\varphi| > \delta} f(\theta - \varphi) \cot(\varphi/2) \, d\varphi .$$

We claim that whenever $f \in L^p$, then $T_\delta f \to \tilde{f}$ in L^p.

Now, by **(2)**, if we define $Q_r * f = v(re^{i\theta})$, then

$$\sup_{r<1} \int_0^{2\pi} |v(r^{i\theta})|^p \, d\theta < \infty .$$

Since v is harmonic, we have, by our theory for Poisson integrals, that $v = P_r * g$ for some $g \in L^p(S)$. Also $v \to g$ in L^p. This function g is defined to be $\tilde{f}$. Since g is a limit of $v(re^{i\theta})$, we have

$$\|g\|_{L^p}^p \le \sup_{r<1} \int_0^{2\pi} |v(re^{i\theta})|^p \, d\theta \le A_p \|f\|_p^p .$$

Now we must see that

$$Q_r(f) - T_\delta(f) \to 0 \text{ in } L^p \text{ norm, with } \delta = 1 - r, \ 1 < p < \infty .$$

This will be clear once we see that the kernels are uniformly L^1 bounded. This follows of course from the estimate

$$|K_r| \le \frac{C\delta}{\varphi^2 + \delta^2}$$

that we proved just before the statement of Riesz's theorem.

Fig. 4.1 Counterexample for
$p = 1, \infty$

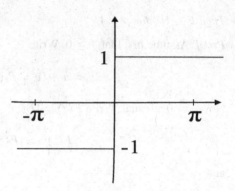

It has already been indicated how **(3)** relates to **(1)** and **(2)**, so we are done. □

Remark 4.1.9 The theory, as we have noted, breaks down for $p = 1$, $p = \infty$. Consider the function $f(\theta)$ shown in Fig. 4.1.

This f is in L^∞, but $Q_r * f$ has essentially no cancelation since Q_r and f are both odd. So

$$|Q_r * f(\theta)| \sim \log(1/|\theta|).$$

This is essentially the worst possible unboundedness. By duality, L^1 must fail in the same way.

Theorem 4.1.10 (The Case $p = 1$) *The Hilbert transform is not defined on all of L^1.*

More precisely, let

$$f_\rho(\theta) = \frac{1}{2\pi} P_\rho(\theta).$$

These f_ρ are uniformly in L^1, in fact $\|f_\rho\|_{L^1} \le 1$. Let us consider

$$Q_{r\rho} = Q_r * f_\rho.$$

If it held that $\|Q_r * f\|_1 \le A\|f\|_1$, then we would have $\|Q_{r\rho}\|_1 \le A\|f_\rho\|_1 \le A$. That is, $c \log(1/(1 - r)) \le A$, and that fails as $r \to 1$.

There are some positive results for H^p spaces, $p \le 1$. The theory divides into two cases—$p = 1$ and $p < 1$. Unlike L^p for $p < 1$, H^p has lots of nontrivial linear functionals. For example, point evaluation on the disc is one such, or point evaluation with any number of derivatives.

Theorem 4.1.11 *Let $f \in L^1(S)$. Write*

$$F(re^{i\theta}) = P_r(f) + iQ_r(f).$$

Then $F \in H^p$ for $p < 1$.

Proof Assume first that $f \geq 0$. Write

$$F = u + iv = P_r(f) + iQ_r(f).$$

Now $u > 0$ as usual. Write $F(z) = |F(z)| \cdot e^{i\Psi(z)}$. Then $|\Psi| < \pi/2$. So

$$F(z)^p = |F(z)|^p e^{ip\Psi(z)}$$

and

$$|p\Psi(z)| < \frac{\pi}{2} \cdot p \text{ if } p < 1.$$

So

$$\frac{1}{2\pi} \int_0^{2\pi} F(re^{i\theta})^p \, d\theta = F(0)^p = a_0^p = \left(\frac{\|f\|_1}{2\pi}\right)^p,$$

that is,

$$\frac{1}{2\pi} \int_0^{2\pi} |F(re^{i\theta})|^p e^{ip\Psi} \, d\theta = a_0^p = \left(\frac{\|f\|_1}{2\pi}\right)^p$$

or

$$\frac{1}{2\pi} \int_0^{2\pi} |F(re^{i\theta})|^p (\text{Re}\,(e^{ip\Psi})) \, d\theta \leq \left(\frac{\|f\|_1}{2\pi}\right)^p$$

so that

$$\frac{1}{2\pi} \int_0^{2\pi} |F(re^{i\theta})|^p \, d\theta \leq \frac{1}{\cos(\pi p/2)} \cdot \frac{\|f\|_1^p}{(2\pi)^p}.$$

That completes the proof.   □

In our study of H^p for $p < 1$, the main idea is that $F \in H^p$, $F \not\equiv 0$, does not have "too many" zeros. In fact:

Theorem 4.1.12 (Jensen) *If $F \in H^p$, let $a_0, a_1, \ldots$ be its zeros in D, counted according to multiplicity, and suppose that $|a_j| \leq |a_{j+1}|$ for each j. Then*

$$\sum_{j=0}^{\infty} (1 - |a_j|) < \infty.$$

Proof We will prove this later. □

Remark 4.1.13 Suppose that $F \in H(D)$, $F(z) \neq 0$ for all $z \in D$. Then $\log |F| \in h(D)$. In fact, $F \in H(D)$ implies that $\log |F| \in h(D)$ if and only if F is never 0 on D.

Proof If $F = u + iv$, F never 0, then $F = e^G$. Write $G = u + iv$. Then $\log |F| = u \in h(D)$.

Conversely, if $F(a) = 0$ for some a, write $F(z) = (z - a)^k \cdot G(z)$, $G(a) \neq 0$. So

$$\log |F| = k \log |z - a| + \log |G(z)| \,.$$

The proof of our remark will follow from the following lemma.

Lemma 4.1.14 Suppose that $F \in H(\mathcal{R})$ and F vanishes at a_k (distinct and finitely many) with multiplicity n_k. Then

$$\triangle \log |F| = 2\pi \sum n_k \delta(z - a_k) \,.$$

Proof of the Lemma Let φ be a test function. We must show that

$$\int_{\mathcal{R}} \log |F| (\triangle \varphi) \, dx dy = 2\pi \sum n_k \varphi(a_k) \,.$$

By Green's theorem, if $B_k = B(a_k, r_k)$ is a small disc centered at a_k with radius r_k, then we have

$$\int_{\mathcal{R}} \log |F| (\triangle \varphi) \, dx dy = \lim_{r_k \to 0} \int_{\mathcal{R} \setminus B_k} \log |F| (\triangle \varphi) \, dx dy$$

$$= \lim_{r_k \to 0} \int_{\partial \mathcal{R}} \log |F| \frac{\partial \varphi}{\partial \eta} \, d\sigma - \lim_{r_k \to 0} \int_{\partial \mathcal{R}} \frac{\partial \log |F|}{\partial \eta} \varphi \, d\sigma$$

$$+ \sum_k \lim_{r_k \to 0} \int_{\partial B_k} \log |F| \frac{\partial \varphi}{\partial \eta_k} \, d\sigma_k - \sum_k \lim_{r_k \to 0} \int_{\partial B_k} \frac{\partial \log |F|}{\partial \eta_k} \varphi \, d\sigma_k$$

$$+ \lim_{r_k \to 0} \int_{\mathcal{R} \setminus B_k} \triangle (\log |F|) \cdot \varphi \, dx dy \,.$$

Now φ is compactly supported in $\mathcal{R}$ and $\log |F|$ is harmonic away from the a_k, so the first, second, and fifth terms vanish. Also, each summand in the third term has magnitude $\sim C |r_k \log r_k| \to 0$. So what we have derived is

$$\int_{R} \log |F| (\triangle \varphi) \, dx dy = - \sum_k \lim_{r_k \to 0} \int_{\partial B_k} \frac{\partial \log |F|}{\partial \eta_k} \cdot \varphi \, d\sigma_k$$

$$= -\sum_k \lim_{r_k \to 0} \int_{\partial B_k} \left[\frac{\partial}{\partial \eta_k} \log |z - a_k|^{n_k} \right] \varphi(z) \, d\sigma_k(z)$$

$$- \sum \lim_{r_k \to 0} \int_{\partial B_k} \left[\frac{\partial}{\partial \eta_k} \log |g_k| \right] \varphi(z) \, d\sigma_k(z), \qquad (\star)$$

where $g_k \equiv F/(z - a_k)^{n_k}$. Since g vanishes to order n_k at a_k, g_k does not vanish in a sufficiently small neighborhood of a_k. So $\partial \log |g_k|/\partial \eta_k$ is smooth near a_k, and the last term on the right of $(\star)$ tends to 0. Of course, $\partial \log |z - a_k|/\partial \eta_k = 1/|z - a_k|$. So the first sum of integrals tends to $\sum 2\pi \delta(z - a_k) \cdot n_k$ as desired. This proves the lemma as well as the second half of the remark. We also now prove:

Lemma 4.1.15 *We have that*

$$\Delta(|F|)^p = p^2 |F|^{p-2} \cdot |F'|^2$$

even if F has zeros.

Proof We wish to show that the identity holds in the sense of distributions. So we need to see that

$$\int |F|^p \Delta \varphi \, dx dy = \int p^2 |F|^{p-2} |F'|^2 \varphi \, dx dy$$

for a test function φ. Near a_k, $F = (z \cdot a_k)^{n_k} g_k(z)$ with a_k, g_k as above. So

$$p^2 |F|^{p-2} |F'|^2 \approx |z - a_k|^{n_k(p-2)+(n_k-1)\cdot 2} \approx |z - a_k|^{n_k p - 2}.$$

Since $n_k \geq 1$ and $p > 0$, this function is locally integrable on $\mathbb{C}^1 \equiv \mathbb{R}^2$. Now we use Green's theorem. Let $\widetilde{\mathcal{R}} = \mathcal{R} \setminus \cup B(a_k, \epsilon)$. Let $u = |F|^p$ and $v = \varphi$. We now proceed exactly as in the proof of the preceding lemma. Since

$$\int_{\widetilde{\mathcal{R}}} u \Delta v - v \Delta u = \int_{\partial \mathcal{R}} u \frac{\partial v}{\partial \eta} - v \frac{\partial u}{\partial \eta},$$

we obtain

$$\int_{\widetilde{\mathcal{R}}} |F|^p \Delta \varphi - \int_{\widetilde{\mathcal{R}}} p^2 |F|^{p-2} |F'|^2 \varphi \, dx dy = \int_{\partial \widetilde{\mathcal{R}}} (\text{junk}).$$

It can be checked that, as $\epsilon \to 0$, the terms on the right tend to 0, whereas the terms on the left tend to

$$\int_{\mathcal{R}} |F|^p \Delta \varphi - \int_{\mathcal{R}} p^2 |F|^{p-2} |F'|^2 \varphi \, dx dy.$$

This completes the proof (Fig. 4.2).   □

Fig. 4.2 The domain with neighborhoods of the a_k deleted

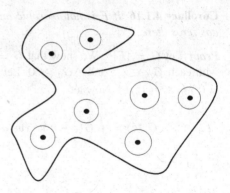

Observations

(1) The second lemma is true for all $p > 0$.
(2) The first lemma is the limit in the appropriate sense of the second lemma because

$$\lim_{p \to 0} \frac{t^p - 1}{p} = \frac{d}{dp} t^p \bigg|_{p=0} = \log t.$$

That is, since

$$\int_{\mathcal{R}} |F|^p \Delta \varphi = \int_{\mathcal{R}} p^2 |F|^{p-2} |F'|^2 \varphi,$$

and since $\int_{\mathcal{R}} \Delta \varphi = 0$, we have

$$\int_{\mathcal{R}} \frac{|F|^p - 1}{p} \Delta \varphi = \int_{\mathcal{R}} p |F|^{p-2} |F'|^2 \varphi.$$

As $p \to 0$, the left side tends to $\int_{\mathcal{R}} (\log |F|) \varphi$.

If we suppose that φ is supported very near to a_k, a_k a zero of F of multiplicity n_k, then the right side

$$\sim \int_{B(a_k, \epsilon)} p |z - a_k|^{n_k(p-2)} n_k^2 |z - a_k|^{(n_k-1)2} \cdot \varphi(a_k)$$

$$= n_k^2 p \left[\int_{B(a_k, \epsilon)} |z - a_k|^{n_k p - 2} \right] \varphi(a_k)$$

$$= 2\pi n_k \epsilon^{n_k p} \varphi(a_k) \to 2\pi n_k \varphi(a_k).$$

This can be made precise using Green's theorem. A corollary of the second lemma is this:

Corollary 4.1.16 *If F is holomorphic on $\mathcal{R}$, then $|F|^p$ is subharmonic even if F has zeros there.*

Proof Let $G = |F|^p$. We show that $G = \lim G_n$, uniformly on compact subsets, with each $G_n \in C^2$ and $\Delta G_n \geq 0$. Let $\varphi \in C_0^\infty(D)$, $\varphi \geq 0$, $\int \varphi = 1$. Define $\varphi_\epsilon(z) = \epsilon^{-2}\varphi(z/\epsilon)$. Now let

$$G_\epsilon(z) = \int G(z - w)\varphi_\epsilon(w)\,dw = \int G(w)\varphi_\epsilon(z - w)\,dw\,.$$

Let

$$\mathcal{R}_\epsilon = \{z \in \mathcal{R} : d(z, {}^c\mathcal{R}) > \epsilon\}\,.$$

Think of G_ϵ as being defined only on $\mathcal{R}_\epsilon$. Clearly, $G_\epsilon \in C^\infty$ and $\Delta G_\epsilon \geq 0$ by the second lemma. The convergence is clear.   □

Corollary 4.1.17 *If $F \in H(D)$, $p > 0$, then*

$$\psi(r) = \int_0^{2\pi} |F(re^{i\theta})|^p\,d\theta$$

is an increasing function of r.

Proof This is true if we replace the integrand $|F|^p$ by *any* subharmonic function g. For let $0 < r_1 < r_2 < 1$. Let $f_{r_2}(\theta) = g(r_2, \theta)$ and $f_{r_1}(\theta) = g(r_1, \theta)$. Let g_j^* be the solution of the Dirichlet problem on $\overline{B}(0, r_j)$ with boundary data f_{r_j}. Then

$$\int_0^{2\pi} g(r_1, \theta)\,d\theta \leq \int_0^{2\pi} g_2^*(r_1, \theta)\,d\theta$$

(since $g_2^* = g$ on $|z| = r_2$ and g_2^* is harmonic, g subharmonic). But this last equals

$$2\pi g_2^*(0) = \int_0^{2\pi} g_2^*(r_2, \theta)\,d\theta = \int_0^{2\pi} g(r_2, \theta)\,d\theta$$

as desired.   □

4.2 Blaschke Products

Let

$$b(z, a) = \frac{z - a}{z - \overline{a}^{-1}} \cdot \frac{1}{|a|}\,, \quad |a| < 1\,, \ a \neq 0\,.$$

Recall that all linear fractional transformations on D are given by

$$z \mapsto e^{i\beta} \left(\frac{z - \alpha}{1 - \overline{\alpha}z} \right) \ , \ \beta \text{ real, } |\alpha| < 1 \,.$$

Note that $b(a, a) = 0$.

Theorem 4.2.1 *Suppose that $F \in H^p(D)$, $0 < p \leq \infty$. Further assume that $F \not\equiv 0$. Then there exists a sequence of $a_j \in D$ such that F vanishes precisely at the a_j. If the a_js are listed counting multiplicities and if F vanishes at the origin to order k, then:*

(1) $\sum (1 - |a_j|) < \infty.$
(2) $z^k \prod_{j=1}^{\infty} b(z, a_j) \equiv B(z)$ *converges and is in $H(D)$. Here k is the order of the zero at $z = 0$.*

One has $|B(z)| \leq 1$ on D and $F(z) = G(z) \cdot B(z)$ with G nonvanishing and $\|G\|_{H^p} = \|F\|_{H^p}$. The convergence of the product in **(2)** *is uniform on compact subsets of D.*

Proof The main tool in the proof of this theorem is *Jensen's formula*. Recall that $F \in H(D)$ and F nonvanishing implies that

$$\frac{1}{2\pi} \int_0^{2\pi} \log |F(re^{i\theta})| \, d\theta = \log |F(0)| \,.$$

However, if F has zeros in D, say $a_1, a_2, \ldots, a_{\ell_r}$ are the zeros of F in a closed subdisc of radius r, no $a_j = 0$, and if F vanishes to order k at 0, then

$$\frac{1}{2\pi} \int_0^{2\pi} \log |F(re^{i\theta})| \, d\theta = \log \left| \left(\frac{F(z)}{z^k} \right)_{z=0} \right| + \log \prod_j \frac{r}{|a_j|} + k \log r \,.$$

We give two proofs of this formula.

Proof I Assume for convenience that $|a_j| \neq r$ for all j. We write

$$F(z) = z^k \prod b \left(\frac{z}{r}, \frac{a_j}{r} \right) F_1(z) \,,$$

where of course F_1 is defined by this formula and F_1 has no zeros in D, $F_1 \in H(D)$. Then

$$\frac{1}{2\pi} \int_0^{2\pi} \log |F_1(re^{i\theta})| d\theta = \log |F_1(0)|$$

$$= \log \left| \left(\frac{F(z)}{z^k} \right)_{z=0} \right| - \sum_{j=1}^{\ell_r} \log |b(0/r, a_j/r)|$$

$$= \log \left| \left(\frac{F(z)}{z^k} \right)_{z=0} \right| + \sum_{j=1}^{\ell_r} \log \frac{r}{|a_j|} \, .$$

But since

$$|b(z,a)| = \left| \frac{z-a}{z - \overline{a}^{-1}} \right| \cdot \frac{1}{|a|} = \left| \frac{z-a}{\overline{a}z - 1} \right| = \left| \frac{z-a}{\overline{z} - \overline{a}} \right| = 1$$

when $|z| = 1$, we have that

$$\left| b \left(\frac{z}{r}, \frac{a_j}{r} \right) \right| = 1$$

when $|z| = r$ so that

$$\frac{|F(z)|}{r^k} = |F_1(z)|$$

when $|z| = r$. Thus

$$\frac{1}{2\pi} \int_0^{2\pi} \log |F(re^{i\theta})| \, d\theta = k \log r + \frac{1}{2\pi} \int_0^{2\pi} \log |F_1(re^{i\theta})| \, d\theta$$

$$= k \log r + \log \left| \left(\frac{F(z)}{z^k} \right)_{z=0} \right| + \sum_{j=1}^{\ell_r} \log \frac{r}{|a_j|}$$

as desired.   □

Proof II Let G be given on D. If $\triangle G = 0$, then of course

$$\frac{1}{2\pi} \int_0^{2\pi} G(re^{i\theta}) \, d\theta = G(0) \, .$$

But if $\triangle G = \mu \neq 0$, then there is an analogue. Let

$$\mathcal{J}(r) = \int_0^{2\pi} G(re^{i\theta}), \, d\theta \, .$$

By Green's theorem,

$$r \cdot \frac{d}{dr} \mathcal{J}(r) = \int_{|z| \le r} \triangle G = \int_{|z| \le r} \mu \, dx dy \, .$$

Since we know $\triangle \log |F|$ explicitly, we can use this identity to compute Jensen's formula. If a_j are the distinct zeros of F with order n_j and F has a zero of order k

at $z = 0$, then we let

$$G(z) = \log \left| \frac{F(z)}{z^k} \right| .$$

Thus

$$\mathcal{J}(r) = \int_0^{2\pi} \log \left| \frac{F(re^{i\theta})}{r^k} \right| d\theta$$

and

$$r \frac{d}{dr} \mathcal{J}(r) = \int_{|z| \leq r} 2\pi \sum n_j \delta(z - a_j) \, dx \, dy .$$

In other words,

$$\frac{d}{dr} \mathcal{J}(r) = \frac{1}{r} 2\pi \chi_r ,$$

where $\chi_r = \sum_{|a_j| \leq r} n_j$. Integrating from 0 to R gives

$$\mathcal{J}(R) - \mathcal{J}(0) = 2\pi \sum_{|a_j| \leq R} n_j \log \frac{R}{|a_j|}$$

or

$$\int_0^{2\pi} \log |F(Re^{i\theta})| \, d\theta = 2\pi k \log R + 2\pi \sum_{|a_j| \leq R} n_j \log \frac{R}{|a_j|} + 2\pi \log \left| \left(\frac{F(z)}{z^k} \right)_{z=0} \right| .$$

This is Jensen's formula, modulo notation. $\square$

We now require the following result.

Lemma 4.2.2 *Let $F \in H(D)$ and $\int_0^{2\pi} \log |F(re^{i\theta})| \, d\theta \leq M < \infty, 0 < r < 1$. If $\{a_j\}$ is a list of the zeros of f, counting multiplicities, then $\sum (1 - |a_j|) < \infty$.*

Proof We have that

$$\left| \sum_{|a_j| \leq r} \log \frac{r}{|a_j|} \right| = \left| \frac{1}{2\pi} \int_0^{2\pi} \log |F(re^{i\theta})| \, d\theta - \log \left| \left(\frac{F(z)}{z^k} \right)_{z=0} \right| - k \log r \right|$$

$$\leq M' < \infty$$

by Jensen's inequality.

Hence, $\sum_j \log 1/|a_j| < \infty$ by monotonicity. Therefore, $\prod_j 1/|a_j| < \infty$ and so $\sum_j (1 - |a_j|) < \infty$. $\qquad\square$

Now define

$$\log^+ u = \begin{cases} \log u & \text{if} \qquad u \geq 1 \\ 0 & \text{if } 0 < u < 1. \end{cases}$$

For any $p > 0$, $u^p > p \log^+ u$ if $u > 0$. Thus $F \in H^p$ implies

$$\int_0^{2\pi} |F(re^{i\theta})|^p \, d\theta \leq \|F\|_p^p,$$

and this in turn yields

$$\int \log^+ |F(re^{i\theta})| \, d\theta \leq M_p < \infty.$$

Hence

$$\int \log |F(r^{i\theta})| \, d\theta \leq M < \infty,$$

so

$$\sum (1 - |a_j|) < \infty.$$

We wish to see that if $F \in H^p$ and $\{a_j\}$ is a list of its zeros (with F vanishing to order k at 0), then

$$B(z) = z^k \prod_{j=1}^{\infty} \left(\frac{z - a_j}{z - \overline{a}_j^{-1}} \right) \cdot \frac{1}{|a_j|}$$

converges uniformly on compact subsets of D. We already know that $\prod_j 1/|a_j|$ converges, so it suffices to see that

$$\prod_{j=1}^{\infty} \left(\frac{z - a_j}{z - \overline{a}_j^{-1}} \right)$$

converges uniformly on $\{|z| \leq r_0\} \subseteq D, r_0 < 1$. That is, we must see that

$$\sum_j \left| 1 - \frac{z - a_j}{z - \overline{a}_j^{-1}} \right|$$

converges uniformly for $|z| \le r_0$.

But

$$\left| 1 - \frac{z - a_j}{z - \overline{a}_j^{-1}} \right| = \left| \frac{a_j - \overline{a}_j^{-1}}{z - \overline{a}_j^{-1}} \right|$$

$$= \left| \frac{1 - |a_j|^2}{z\overline{a}_j - 1} \right|$$

$$\le \frac{2|1 - |a_j||}{1 - |z|}$$

$$\le \frac{2(1 - |a_j|)}{1 - r_0},$$

which sums.

Now define $G(z) = F(z)/B(z)$. Clearly, $\|F\|_{H^p} \le \|G\|_{H^p}$ since $|B| \le 1$. We claim that the reverse inequality also holds. Let

$$G_N(z) = \frac{F(z)}{B_N(z)},$$

where

$$B_N(z) = z^k \prod_{j=1}^{N} b_j(z).$$

For N fixed, $\int_0^{2\pi} |G_N(re^{i\theta})|^p \, d\theta$ is an increasing function of r. Therefore,

$$\int_0^{2\pi} |G_N(re^{i\theta})|^p \, d\theta \le \sup_{s<1} \int_0^{2\pi} |F(se^{i\theta})|^p \, d\theta = \|F\|_p^p$$

since $B_N(se^{i\theta}) \to 1$ uniformly as $s \to 1$. Letting $N \to \infty$, we see that

$$\int_0^{2\pi} |G(re^{i\theta})|^p \, d\theta \le \|F\|_p^p$$

for all r.

Theorem 4.2.3 *Let $F \in H^p, 0 < p \le \infty$. Then:*

1. $\lim_{r \to 1} F(re^{i\theta}) \equiv F(e^{i\theta})$ *exists for almost every θ. The limit exists in the nontangential sense as well.*
2. $\int_0^{2\pi} |F(re^{i\theta}) - F(e^{i\theta})|^p \, d\theta \to 0$ *as $r \to 1$.*

3. $\int_0^{2\pi} \left(\sup_{z \in T_\theta^\alpha} |F(z)| \right)^p d\theta \leq C_p^\alpha \|F\|_p^p$, *where T_θ^α is the cone of aperture α, vertex θ.*

Remark We had this result for F merely harmonic, $p > 1$. When $p = 1$, we had weakened versions of **1** and **2**, but **3** failed.

Proof of the Theorem Let $F \in H^p$, and let $F = G \cdot B$ be its Blaschke factorization. Then we write

$$F = G(B - 1) + G \equiv F_1 + F_2,$$

where the F_j are zero-free and each $F_j \in H^p$.

Then each $F_j^{p/2} \in H^2$ so we know that each $F_j^{p/2}$ has nontangential boundary values almost everywhere. Hence, so does each F_j hence so does F. This proves **1**. Moreover,

$$\int_0^{2\pi} \left(\sup_{z \in T_\theta} |F_j^{p/2}(z)| \right)^2 d\theta \leq C_2 \|F_j^{p/2}\|_2^2$$

or

$$\int_0^{2\pi} \left(\sup_{z \in T_\theta} |F_j(z)| \right)^p d\theta \leq C_2 \|F_j\|_p^p$$

so

$$\int_0^{2\pi} \left(\sup_{z \in T_\theta} |F(z)| \right)^p d\theta \leq 4C_2 \|F\|_p^p.$$

That proves **3**.

For **(2)**, suppose first that $p > 1/2$. Then

$$\int |F_j(re^{i\theta}) - F_j(e^{i\theta})|^p d\theta = \int |F_j^{1/2}(re^{i\theta}) - F_j^{1/2}(e^{i\theta})|^p \cdot |F_j^{1/2}(re^{i\theta}) + F_j^{1/2}(e^{i\theta})|^p d\theta$$

$$\leq \int |F_j^{1/2}(re^{i\theta}) - F_j^{1/2}(e^{i\theta})|^{2p} d\theta^{1/2} \times$$

$$\int |F_j^{1/2}(re^{i\theta}) + F_j^{1/2}(e^{i\theta})|^{2p} d\theta^{1/2}.$$

Now $F_j^{1/2} \in H^{2p}$ and $2p > 1$, so the first term tends to 0 by the old theory and the second term is bounded by $2\|F\|_{H^p}$.

This takes care of the case $p > 2^{-1}$. Inductively, once we know the result for $p > 2^{-k}$, we may obtain that for $p > 2^{-k-1}$ in the same way.   □

Corollary 4.2.4 *We have that $F \in H^1$ if and only if there exists f real-valued and in L^1 such that $\widetilde{f} \in L^1$ and there exists a constant b_0 such that*

$$F(re^{i\theta}) = P_r * f(e^{i\theta}) + iP_r * \widetilde{f}(e^{i\theta}) + ib_0$$

and

$$\|F\|_{H^1} \approx \|f\|_{L^1} + \|\widetilde{f}\|_{L^1} + |b_0|.$$

Proof By the theorem, part **2**, convergence of F to its boundary function is in L^1 norm so Re F is the Poisson integral of an L^1 function and so is Im F. □

Remark 4.2.5 There is an ambiguity in this corollary. Namely, what do we mean by saying that $\widetilde{f} \in L^1$? We could mean two things:

(1) That $f \sim \sum a_n e^{in\theta}$, and $\sum -i \operatorname{sgn} n \cdot a_n e^{in\theta}$ is the Fourier series of an L^1 function.
(2) That $\lim_{r \to 1} Q_r * f \equiv \widetilde{f} \in L^1$ or

$$\lim_{\delta \to 0} \frac{1}{2\pi} \int_{|\varphi| \geq \delta} f(\theta - \varphi) \cot \frac{\varphi}{2} \, d\varphi \equiv \widetilde{f} \in L^1.$$

It is not difficult to see that **(2)** $\Rightarrow$ **(1)**. But **(1)** $\Rightarrow$ **(2)** is more subtle. This is resolved by using Bok's integral (cf. [ZYG]).

Corollary 4.2.6 *Suppose that $d\mu$ is a finite Radon measure, $d\mu \sim \sum_{n \geq 0} c_n e^{in\theta}$. Then $d\mu$ is absolutely continuous with respect to Lebesgue measure.*

Proof Each $|c_n| \leq \|d\mu\|$. So $F(z) = \sum c_n r^n e^{in\theta} = \sum c_n z^n$ is holomorphic. We have that $F(re^{i\theta}) = P_r * d\mu$ so

$$\int_0^{2\pi} |F(re^{i\theta})| \, d\theta \leq \frac{1}{2\pi} \|P_r\|_1 \|d\mu\|$$

for all r. Thus $F \in H^1$. Therefore, $F(re^{i\theta})$ converges to some L^1 function f as $r \to 1$. But also $F(re^{i\theta}) \to d\mu$ in the weak-$*$ topology. Therefore, $d\mu = f \in L^1$. □

Now we know that $F \in H^p$, $p > 1$, if and only if there exists an $f \in L^p$ such that $\|F\|_{H^p} \approx \|f\|_{L^p}$ and $f = \operatorname{Re} F$. For $p = 1$, this no longer holds, and one needs the Hilbert transform, or a maximal function, or a variety of other substitutes. For $p < 1$, it is not clear what is needed.

We now discuss the "largest" class of functions that contains all the H^p functions and for which the above methods work. We refer, of course, to the Nevanlinna class N. Here,

$$N = \{F \in H(D) : \sup_{r<1} \int_0^{2\pi} \log^+ |F(re^{i\theta})| \, d\theta < \infty\}.$$

Theorem 4.2.7 *If $p > 0$, then $H^p \subseteq N$.*

Proof This follows since $p \log^+ u \le u^p$. $\hfill\square$

Theorem 4.2.8 *If $F \in N$, then $F = GB$, B a Blaschke product, and $G \in H(D)$ has no zeros, $G \in N$.*

Theorem 4.2.9 *If $F \in N$, then $\lim_{r \to 1} F(re^{i\theta})$ exists almost everywhere (and also nontangentially).*

The space N is essentially best possible, because the result is false for the class $\text{Log}^{1-\epsilon}$ and also for the class $\text{Log}/(\text{LogLog})$.

To see why the above are true, note that $F \in N$ implies that $\int_0^{2\pi} \log |F(re^{i\theta})| \, d\theta$ remains bounded above so, by the old argument, the zeros of F are sparse, and there exists a Blaschke product B with $F = GB$. We need to see that $G \in N$. The key fact is that

$$\int \log^+ |F(re^{i\theta})| \, d\theta$$

is increasing in r, $0 < r < 1$. This would follow from the subharmonicity of $\log^+ |F|$, but we never bothered to define subharmonicity for discontinuous functions. We will show that if $F \in H(\overline{D})$, then

$$\int \log^+ |F(re^{i\theta})| \, d\theta \le \int \log^+ |F(e^{i\theta})| \, d\theta$$

for all $0 < r < 1$. This will certainly suffice.

Now $|F|^p$ is subharmonic in the closed disc so

$$|F(re^{i\theta})|^p \le \frac{1}{2\pi} \int_0^{2\pi} P_r(\theta - \varphi) |F(e^{i\varphi})|^p \, d\varphi.$$

Hence,

$$\frac{|F(re^{i\theta})|^p - 1}{p} \le \frac{1}{2\pi} \int_0^{2\pi} P_r(\theta - \varphi) \left(\frac{|F(e^{i\varphi})|^p - 1}{p} \right) d\varphi.$$

Let $p \to 0$. So

$$\log |F(re^{i\theta})| \le \frac{1}{2\pi} \int_0^{2\pi} P_r(\theta - \varphi) \log |F(e^{i\varphi})| \, d\varphi.$$

This passing to the limit is justified because F vanishes only finitely many places on ∂D, and we have pointwise convergence and trivial domination. So

$$\log^+ |F(re^{i\theta})|\, d\theta \le \frac{1}{2\pi} \int_0^{2\pi} P_r(\theta - \varphi) \log^+ |F(\epsilon^{i\varphi})|\, d\varphi .$$

Now integrate over θ so

$$\frac{1}{2\pi} \int_0^{2\pi} \log^+ |F(re^{i\theta})|\, d\theta \le \frac{1}{2\pi} \int_0^{2\pi} \log^+ |F(e^{i\varphi})|\, d\varphi .$$

This proves the claim, and $G \in N$ follows as usual. Also G does not vanish on D. So $G = e^H = e^{u+iv}$, some $H \in H(D)$.

Now

$$\sup_r \int_0^{2\pi} \log^+ |G(re^{i\theta})|\, d\theta < \infty$$

if and only if

$$\sup_r \int_0^{2\pi} u^+(re^{i\theta})\, d\theta < \infty .$$

But

$$\int_0^{2\pi} u(re^{i\theta})\, d\theta = u(0)$$

is a constant, so

$$\sup_r \int u^-(re^{i\theta})\, d\theta$$

is finite. Hence,

$$\sup_r \int_0^{2\pi} |u|(re^{i\theta})\, d\theta < \infty .$$

Therefore, u is the Poisson integral of some finite Radon measure $d\mu$. It follows easily that $u + iv \in H^p$ for all $p < 1$, so $H = u + iv$ has boundary values almost everywhere; hence so does G, hence F. $\qquad\square$

4.3 Passage from D to $\mathbb{R}^2_+$

What happens when we replace D by $\mathbb{R}^2_+$? Let

$$P_y(x) = \frac{y}{\pi(x^2 + y^2)} \ , \quad Q_y(x) = \frac{x}{\pi(x^2 + y^2)} \ .$$

These are essentially the real and imaginary parts of the boundary values of the Cauchy kernel. We say that $F \in H^p(\mathbb{R}_+^2)$ if $F \in H(\mathbb{R}_+^2)$ and $\sup_{y>0} \int |F(x + iy)|^p \, dx < \infty$.

Theorem 4.3.1 *We have the following:*

1. *Let* $f \in L^p(\mathbb{R}^1)$, $1 < p < \infty$, *and let*

$$T_\delta f(x) = \frac{1}{\pi} \int_{|t| \geq \delta} \frac{f(x - t)}{t} \, dt \ .$$

Then

$$\lim_{\delta \to 0} T_\delta f = Tf = \widetilde{f}$$

exists almost everywhere and in L^p norm. Also $f \mapsto \widetilde{f}$ is a bounded operator on L^p.

2. *The function $F \in H^p$, $1 < p < \infty$, if and only if there exists $f \in L^p$, real-valued, such that*

$$F = P_y(f) + i P_y(\widetilde{f}) = P_y(f) + i Q_y(f)$$

with $\|F\|_{H^p} \approx \|f\|_{L^p}$.

Proof The proof is nearly identical to that on the disc, and we omit the details.   □

The easiest way to obtain Blaschke products on $\mathbb{R}_+^2$ is via the Cayley transform.

$$\gamma : z \mapsto w(z) = \frac{1}{i}\left(\frac{z - 1}{z + 1}\right) \ .$$

The inverse of this map is

$$z = \frac{i - w}{i + w} \ .$$

Refer to Fig. 4.3.

If $F \in H^p$, $p > 0$, and the $\{a_j\}$ are the zeros of F, counting multiplicities, then we have

$$\sum_j \frac{\mathrm{Im}\,(a_j)}{|a_j + 1|^2} < \infty \ .$$

It follows that

Fig. 4.3 The Cayley transform

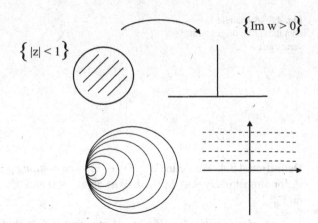

Fig. 4.4 The circles that osculate at -1 correspond to the level lines

$$B(z) = \prod_j b(\gamma^{-1}(z), \gamma^{-1}(a_j))$$

exists, and we can write $F(z) = G(z) \cdot B(z)$. By imitating the proof on the disc, we get almost everywhere boundary values for $F \in H^p$.

It would be nice if the theory on $\mathbb{R}^2_+$ were canonically isomorphic to that on the disc via the Cayley map γ; however, this is not true since the circles centered at 0 do not correspond, under γ, to the lines parallel to the x-axis. Instead the circles tangent at -1 correspond to these lines. This point is illustrated in Fig. 4.4.

Nonetheless, the theories are isomorphic and isometric.

Theorem 4.3.2 *Let* $f : D \to \mathbb{C}$*;* $F : \mathbb{R}^2_+ \to \mathbb{C}$*,* $z \in D$*,* $w \in \mathbb{R}^2_+$ *be related by*

$$F(w) = 2^{1/p}(1 + w)^{-p/2} f(z)$$

with

$$z = \frac{i - w}{i + w}.$$

Then $F \in H^p(\mathbb{R}^2_+)$ *if and only if* $f \in H^p(D)$ *and* $\|F\|_{H^p(\mathbb{R}^2_+)} = \|f\|_{H^p(D)}.$

Proof The proof of this statement requires some work. We omit it.   □

Remark 4.3.3 We would like to formulate local analogues of the global boundary value theorems that we have been discussing. That is, if $E \subseteq \mathbb{R}^1$ is a set of positive measure and $\mathcal{R} \subseteq \mathbb{R}^2_+$, $\partial\mathcal{R} \supseteq E$, and $F \in H(\mathcal{R})$, we would like to have some local condition on F that guarantees that F has boundary values almost everywhere on E.

Fig. 4.5 A triangle
containing the origin with
vertex at 1

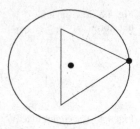

Definition 4.3.4 We define $F : D \to \mathbb{C}$ to be *nontangentially bounded* on $E \subseteq S$ if, for almost every $\theta \in E$, there exist an $\alpha_\theta > 0$ and $M_\theta > 0$ such that $|F| \leq M_\theta$ on $\Gamma_\theta^{\alpha_\theta}$.

A theorem of Privalov–Plessner states that if $F \in h(D)$, then F is nontangentially bounded almost everywhere on $E \subseteq S$ if and only if F has nontangential limits almost everywhere on E.

We sketch the proof of this result using complex function theory. We remark that there is no reason to suspect that, at a given point, nontangential boundedness implies nontangential convergence. In fact, there are counterexamples that oscillate wildly.

Proof of the Privalov–Plessner Theorem Let $u \in h(D)$. Suppose that $E \subseteq S$ is closed without loss of generality. Fix once and for all a triangle T that contains 0 and has 1 as one of its vertices. See Fig. 4.5.

Let T_θ be the same triangle rotated to have vertex at θ. Assume that u is uniformly bounded on all T_θ, $\theta \in E$. We will show that u has nontangential limits almost everywhere on E. In fact, we observe that all possible triangles are contained in one of a countable family (namely those with rational aperture), and all possible bounds are less than or equal to some integer, so we may break E into countably many pieces each of which has the above uniformity property. Since E is closed, ^cE is the disjoint union of open intervals.

On each of these intervals, erect a triangle with the same aperture as T. Consider the shaded region in the figure. It is just

$$\bigcup_{\theta \in E} T_\theta \equiv \mathcal{M} .$$

We claim that $\partial \mathcal{M}$ is rectifiable (Fig. 4.6).

This is clear since the perimeter of each triangle is majorized by a constant C times the length of its base (since they all have the same aperture). So the length of $\partial \mathcal{M}$ is less than or equal to

$$C \cdot (\text{circumference of circle}) \leq 2C\pi .$$

Fig. 4.6 The region $\mathcal{M}$

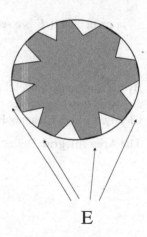

E

Now, by the Riemann mapping theorem, we conformally map $\mathcal{M}$ to the unit disc D. There is a theorem about conformal mapping that guarantees us that conformal maps of the a domain to the disc are as good at the boundary as the boundary of the domain will allow. That is, if the boundary is a Jordan curve, then the mapping extends to a homeomorphism of the boundaries; and if the boundary is rectifiable, then sets of measure 0 coincide under the mapping and the map is conformal at almost every boundary point.

Thus our conformal mapping extends to a homeomorphism of $\partial \mathcal{M}$ to ∂D. And so F is transferred, via the conformal mapping, to a bounded function on D. Hence, this function on D has nontangential almost everywhere boundary values, and since the conformal map extends conformally to $\overline{\mathcal{M}}$, so does F. (One should note that there are some details to be filled in via the familiar density argument.)

Corollary 4.3.5 *If $F \in H(D)$ and F has nontangential limit 0 on a set of positive measures in the boundary, then $F \equiv 0$.*

Proof By the usual uniformization, F can be taken to be bounded on a sawtooth region in D with boundary value 0 on a set of positive measures. Via a conformal map, we get an $F^* \in H^\infty(D)$ with boundary value 0 on a set of positive measures. So $F \equiv 0$.   □

Corollary 4.3.6 *Let $F \in H(D)$, $F = u + iv$, $u, v \in h(D)$ real-valued. Then u has nontangential limits on a set E of positive measures if and only if v does.*

Proof By a uniformizing procedure as above, suppose that $|u| \leq 1$ on an appropriate sawtooth region. Then $1/(u + iv + 2)$ is bounded on this region so, by the theorem, has nontangential limits almost everywhere on E. Hence, so does v. The converse follows by symmetry.   □

Theorem 4.3.7 (Bagemihl–Seidel) *Suppose that $E \subseteq \partial D$, E of first category. Suppose also that $f(re^{i\theta}) \in C(D)$. Then there exists an $F \in H(D)$ such that $F(re^{i\theta}) - f(re^{i\theta}) \to 0$ as $r \to 1$ for all $\theta \in E$.*

In particular, if E is a set of full measures and of first category, let

$$f = PI(\chi_{E \cap [0,\pi]}).$$

By the Bagemihl–Seidel theorem, there is an $F \in H(D)$, $F \not\equiv 0$, such that $F \to 1$ almost everywhere on $E \cap [0, \pi]$ and $F \to 0$ almost everywhere on $E \cap [\pi, 2\pi]$. Clearly, such an F cannot be in $N(D)$.

The Area Integral Define, for $F \in H(D)$,

$$S(F)(\theta) = \left(\iint_{T_\theta} |F'(z)|^2 \, dx dy \right)^{1/2}.$$

This is the Lusin area integral.

If $E \subseteq \partial D$ has positive measure, then F has nontangential limits almost everywhere on E if and only if $|SF| < \infty$ almost everywhere on E. Also, we have

$$|a_0|^2 + \|SF\|_{L^2}^2 \approx \|F\|_{H^2}^2.$$

Proof of the Above Equivalence Now

$$\|S(F)\|_{L^2}^2 = \int_0^{2\pi} |S(F)(\theta)|^2 \, d\theta$$

$$= \int_0^{2\pi} \left(\iint_{T_\theta} |F'(z)|^2 \, dx dy \right) d\theta$$

$$= \iiint \chi(z, \theta) |F'(z)|^2 \, dx dy d\theta,$$

where

$$\chi(z, \theta) = \begin{cases} 1 & \text{if} \quad z \in T_\theta \\ 0 & \text{otherwise}. \end{cases}$$

It is easy to see that

$$\int \chi(z, \theta) \, d\theta \approx 1 - |z|.$$

So

$$\|S(F)\|^2_2 = \iint (1 - |z|)|F'(z)|\, dx dy . \qquad (*)$$

Let

$$F(re^{i\theta}) = \sum a_n r^n e^{in\theta} .$$

So

$$(*) = \int_0^{2\pi} \int_0^1 (1 - r)\left|\sum a_n i n r^{n-1} e^{in\theta}\right|^2 r\, dr d\theta \,(\text{Plancherel})$$

$$= c \int_0^1 \sum |a_n|^2 n^2 r^{2n-1}(1 - r)\, dr .$$

But

$$\int_0^1 r^{2n-1}(1 - r)\, dr = \frac{1}{2n} - \frac{1}{2n+1} = \frac{1}{4n^2} \quad \text{for } n \geq 1 .$$

Hence, the last line

$$\approx \sum_1^\infty |a_n|^2 \approx \|F\|^2_{H^2} - |a_0|^2 .$$

Chapter 5
Hardy Spaces on $\mathbb{R}^n$

5.1 The Poisson Kernel on the Ball

We now study the $\mathbb{R}^n$ theory. Let

$$\mathbb{R}^{n+1}_+ \equiv \{(x, y) : x \in \mathbb{R}^n, y > 0\}.$$

Proposition 5.1.1 *Let*

$$\omega_{n-1} = \text{area of } S^{n-1} = \text{area of } \{x \in \mathbb{R}^n : |x| = 1\}.$$

Then

$$\omega_{n-1} = \frac{2\pi^{n/2}}{\Gamma(n/2)}.$$

Proof Recall that

$$\int_{-\infty}^{\infty} e^{-\pi x^2} \, dx = 1,$$

hence

$$\int_{\mathbb{R}^n} e^{-\pi |x|^2} \, dx = 1.$$

Thus

$$1 = \omega_{n-1} \int_0^{\infty} e^{-\pi r^2} r^n \frac{dr}{r}$$

© The Author(s), under exclusive license to Springer Nature Switzerland AG 2023
S. G. Krantz, *The E. M. Stein Lectures on Hardy Spaces*, Lecture Notes in
Mathematics 2326, https://doi.org/10.1007/978-3-031-21952-8_5

$$= \omega_{n-1} \int_0^\infty e^{-\pi s} s^{n/2} \frac{ds}{2s}$$

$$= \left[\int_0^\infty e^{-s} s^{n/2} \frac{ds}{s} \right] \cdot \frac{1}{\pi^{n/2}} \cdot \frac{\omega_{n-1}}{2}$$

$$= \frac{\Gamma(n/2)}{2\pi^{n/2}} \cdot \omega_{n-1} .$$

$\square$

Proposition 5.1.2

(a) *If $n > 2$, then $|x|^{-n+2}$ is harmonic on $\mathbb{R}^n \setminus \{0\}$. Also $\log |x|$ is harmonic on $\mathbb{R}^2$ for $x \neq 0$.*

(b) *If $f : \mathbb{R}^n \to \mathbb{R}$ is harmonic, then define $F(x) = f(x^*) \cdot |x|^{-n+2}$, where $x^* \equiv x/|x|^2$ for $x \neq 0$. Then F is harmonic. This is Kelvin's principle.*

Proof Verify **(b)** by differentiation. Prove **(a)** by letting $f \equiv 1$ in part **(b)**.

The Poisson Kernel for the Unit Ball in $\mathbb{R}^n$

We seek a function $P(z, \zeta)$ such that, if $f \in C(S^{n-1})$, then

$$F(z) = \int_{|\zeta|=1} P(z, \zeta) f(\zeta) \, d\sigma_\zeta$$

is harmonic on B^n, continuous on $\overline{B^n}$, and $F \Big|_{S^{n-1}} = f$.

Proposition 5.1.3 *We have that*

$$P(z, \zeta) = \frac{1}{\omega_{n-1}} \left(\frac{1 - |z|^2}{|z - \zeta|^n} \right) .$$

How might one guess this formula? It would suffice to find the Green's function for B^n. Recall that the function $G(z, \zeta)$ is uniquely determined by these properties:

(i) $G(z, \zeta)$ is harmonic in ζ for each fixed $z \in B^n$.

(ii) We have that $G(z, \zeta) = 0$ if $|\zeta| = 1$.

(iii) We have

$$G(z, \zeta) = \frac{a_n}{|z - \zeta|^{n-2}} + \text{(something smooth)},$$

where $a_n = (-1)/[\omega_{n-1} \cdot (n - 2)]$ (Fig. 5.1).

Fig. 5.1 The ball and the sphere

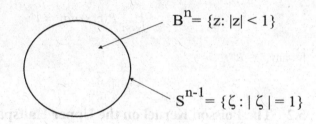

$B^n = \{z : |z| < 1\}$

$S^{n-1} = \{\zeta : |\zeta| = 1\}$

We have indicated earlier how

$$P(z, \zeta) = \frac{\partial}{\partial \eta_\zeta} G(z, \zeta),$$

with $z \in B^n$, $\zeta \in S^{n-1}$. This statement can be checked using Green's formula. To construct G, start with $a_n/|z - \zeta|^{n-2}$. We apply Kelvin's principle to obtain (ii). Namely, the Kelvin transform of $a_n/|z - \zeta|^{n-2}$ is

$$\frac{a_n |\zeta|^{-n+2}}{|z - \zeta^*|^{n-2}}.$$

These two functions agree on S^{n-1}, hence

$$\frac{a_n}{|z - \zeta|^{n-2}} - \frac{a_n |\zeta|^{-n+2}}{|z - \zeta^*|^{n-2}}$$

is harmonic, except at z, and vanishes on S^{n-1}. One can compute

$$\frac{\partial}{\partial \eta_\zeta} G(z, \zeta) = P(z, \zeta) = \frac{1}{\omega_{n-1}} \left(\frac{1 - |z|^2}{|z - \zeta|^n} \right)$$

as claimed.

Since $P(0, \zeta) = 1/\omega_{n-1}$, and since, for fixed ζ, P is harmonic in z (because G is), we have

$$\frac{1}{\omega_{n-1}} = P(0, \zeta) = \frac{1}{\omega_{n-1}} \int_{S^{n-1}} P(rz', \zeta) \, d\sigma_{z'}$$

for any $r < 1$ and z' ranging over S^{n-1}. But $|rz' - \zeta| = |z' - r\zeta|$ so the last line equals

$$\frac{1}{\omega_{n-1}} \int_{S^{n-1}} P(r\zeta, z') \, d\sigma_{z'}.$$

Thus

$$\int_{S^{n-1}} P(z, s)\, d\sigma_s = 1$$

for any $z \in B^n$.

5.2   The Poisson Kernel on the Upper Halfspace

Now we claim that the Poisson kernel on the upper halfspace $\mathbb{R}^{n+1}_+$ is

$$P(z, t) = \frac{c_n y}{(y^2 + |x - t|^2)^{(n+1)/2}}$$

for $t \in \mathbb{R}^n = \partial \mathbb{R}^{n+1}_+$, $z = (x, y) \in \mathbb{R}^{n+1}_+$, and

$$\frac{1}{c_n} = \frac{\omega_n}{2} = \frac{\pi^{(n+1)/2}}{\Gamma((n+1)/2)}.$$

One could again use Green's theorem (and the Green's function) to discover this. In this context, it is easier to use Schwarz reflection. We have

$$G(z, \zeta) = \frac{a_{n+1}}{|z - \zeta|^{n-1}} - \frac{a_{n+1}}{|z - \zeta^*|^{n-1}},$$

where

$$z = (x, y)\ ,\quad \zeta = (\xi, \eta)\ ,\quad \zeta^* = (\xi, -\eta).$$

Hence

$$P(z, \xi) = \frac{\partial}{\partial \eta} G(z, \xi) \ \text{ for } \xi \in \mathbb{R}^n = \partial \mathbb{R}^{n+1}_+.$$

We frequently write

$$P_y(x - t) = P(x + iy, t) = \frac{\partial}{\partial \eta} G(x + iy, t) = \frac{c_n y}{(y^2 + |x - t|^2)^{(n+1)/2}}$$

or

$$P_y(x) = \frac{c_n y}{(|x|^2 + y^2)^{(n+1)/2}} \ \text{ for } x \in \mathbb{R}^n, t > 0.$$

The Fourier analysis derivation of this kernel is

$$P_y(x) \equiv \int_{\mathbb{R}^n} e^{-2\pi y |\xi|} e^{-2\pi i x \cdot \xi} \, d\xi \,.$$

By Fourier inversion, valid after some checking, we have

$$\int P_y(x) \, dx = e^{-2\pi y |\xi|} \bigg|_{\xi=0} = 1 \,.$$

Lemma 5.2.1 *Suppose that u is harmonic on $\mathbb{R}_+^{n+1}$, continuous on $\overline{\mathbb{R}_+^{n+1}}$, bounded, and that $u\big|_{\mathbb{R}^n} = 0$. Then $u \equiv 0$.*

Proof One can prove a better theorem using the Phragmen–Lindelöf technique. Instead we use the Schwarz reflection principle. We define $u(x, y)$ for $y < 0$ by $u(x, y) = -u(x, -y)$. It follows that the extended u is harmonic on all of $\mathbb{R}^{n+1}$. Then $u \equiv 0$ follows by Liouville's theorem. More precisely, choose $x, y \in \mathbb{R}^{n+1}$. Let $r >> |x - y|$. Then

$$|u(x) - u(y)| = \frac{1}{m(B(x, r))} \left| \int_{B(x,r)} u(t) \, dt - \int_{B(y,r)} u(t) \, dt \right|$$

and, as $r \to +\infty$, this clearly tends to 0. $\qquad\qquad\square$

Exercise Using Liouville's principle, show that if u is harmonic on some $\mathbb{R}^n$ and u is of polynomial growth, then u is a polynomial.

Theorem 5.2.2 *Suppose that $u(x, y)$ is harmonic on $\mathbb{R}_+^{n+1}$ and bounded. Then there exists an $f \in L^\infty(\mathbb{R}^n)$ such that $u = PIf$. Also $\|u\|_{L^\infty} \le \|f\|_{L^\infty}$.*

Remark 5.2.3 The converse to this result is clear once we observe that P_y is harmonic (because it is just the derivative of $(|x|^2 + y^2)^{-(n+1)/2}$). Since $\|P_y\|_{L^1} = 1$, we get

$$\|u\|_{L^\infty} \le \|f\|_{L^\infty} \cdot \|P_y\|_{L^1} \le \|f\|_{L^\infty} \,.$$

Proof of the Theorem Let $f_k(x) = u(x, 1/k), k = 1, 2, \dots$. Let $u_k(x, y) = PIf_k$. Look at the functions

$$g_k(x, y) = u(x, y + 1/k) - u_k(x, y) \,.$$

Then

$$g_k \in C(\overline{\mathbb{R}_+^{n+1}}) \cap L^\infty(\mathbb{R}_+^{n+1}) \cap h(\mathbb{R}_+^{n+1})$$

and $g_k(x, 0) = 0$. Thus $g_k \equiv 0$. Now the f_k form a bounded subset of L^∞ so there exists a weak-$*$ convergent subsequence $f_{j_k} \to f \in L^\infty$. Thus

$$u_{j_k} = f_{j_k} * P_y \to f * P_y$$

pointwise. That is, $u(x, y + 1/k) \to f * P_y$. But, by continuity, this says that $u(x, y) = f * P_y$. Now, of course,

$$\|f\|_{L^\infty} \leq \sup \|f_k\|_{L^\infty} \leq \sup_y \|u(x, y)\|_{L^\infty(x)}$$

and, since the reverse is clear, we see that $\|f\|_{L^\infty} = \|u\|_{L^\infty}$.   □

Corollary 5.2.4 *If u is given on $\mathbb{R}_+^{n+1}$, then u is the Poisson integral of an $f \in L^p$, $1 < p \leq \infty$ if and only if u is harmonic and*

$$\sup_y \left(\int_{\mathbb{R}^n} |u(x, y)|^p \, dx \right)^{1/p} < \infty.$$

Moreover,

$$\|f\|_{L^p} = \sup_{y>0} \left(\int |u(x, y)|^p \, dx \right)^{1/p}.$$

Proof The "only if" part is clear as before. The "if" part will follow by a proof similar to that of the theorem once we prove the following lemma. This lemma will enable us to apply the uniqueness theorem on appropriate upper halfspaces.

Lemma 5.2.5 *If*

$$\int |u(x, y)|^p \, dx \leq C,$$

with C independent of y, then $|u(x, y)| \leq C' y^{-n/p}$.

Proof By the mean value property,

$$u(x, y) = \frac{1}{m(B((x, y), y/2))} \int_{B((x,y),y/2)} u(z) \, dz.$$

Hence

$$|u(x, y)|^p \leq \frac{1}{m(B)} \int_B |u(z)|^p \, dz$$

$$\leq \frac{1}{m(B)} \int_{|\operatorname{Im} z - y| < y/2} |u(z)|^p \, dxdy$$

$$\leq \frac{1}{m(B)} Cy$$

$$\leq Cy^{-n}.$$

Thus $|u(x, y)| \leq C y^{-n/p}$. $\square$

So we see that u is bounded on any set $\mathcal{R}_{y_0} = \{(x, y) : y \geq y_0\}$ for $y_0 > 0$. The corollary above now follows by the standard argument. $\square$

Corollary 5.2.6 *The function u is harmonic on $\mathbb{R}_+^{n+1}$ and $\sup_{y>0} \int |u(x, y)| \, dx <$ ∞ if and only if there exists a finite Radon measure $d\mu$ so that $u = P_y * d\mu$. Of course*

$$\sup_{y>0} \int |u(x, y)| \, dx = \|d\mu\| \, .$$

Proof As usual. $\square$

Corollary 5.2.7 *In any of the above, $u(x, y)$ converges almost everywhere as $y \to$ 0, in fact the convergence can be taken to be nontangential. In case $p = 1$, write $d\mu = f(x)dx + d\eta$, where $f \in L^1$ and $d\eta \perp dx$. The $u \to f$ nontangentially.*

Having already used the notion, we now remind the reader what nontangential convergence is. For $x_0 \in \mathbb{R}^n$, define

$$\Gamma_\alpha^h(x_0) = \{(x, y) \in \mathbb{R}_+^{n+1} : |x - x_0| < \alpha y, 0 < y < h\} \, .$$

Definition 5.2.8 A function $u : \mathbb{R}_+^{n+1} \to \mathbb{C}$ is said to be *nontangentially bounded* at $x_0 \in \mathbb{R}^n$ if there exist α, h, depending on x_0, such that $u|_{\Gamma_\alpha^h(x_0)}$ is bounded.

Also, u is said to have a nontangential limit at x_0 if there exists a number ℓ such that, for every $\alpha > 0, h > 0$, one has

$$\lim_{\substack{(x,y) \to (x_0, y) \\ (x,y) \in \Gamma_\alpha^h(x_0)}} = \ell \, .$$

We also define

$$Mf(x) = \sup_{r>0} \frac{1}{\Omega_n r^n} \int_{|y| \leq r} |f(x - y)| \, dy \, ,$$

where Ω_n is the volume of the unit ball in $\mathbb{R}^n$.

Theorem 5.2.9 *We have*

(1) $\sup_{y>0} |u(x, y)| \leq Mf(x)$.
(2) $\sup_{(x,y) \in \Gamma_\alpha(x_0)} |u(x, y)| \leq C_\alpha Mf(x_0)$.
(3) $\lim_{\substack{(x,y) \in \Gamma_\alpha(x_0) \\ (x,y) \to (x_0, 0)}} u(x, y) = f(x_0)$ for all x_0 in the Lebesgue set of f .

Proof of (1) By dilation invariance, it suffices to show that $u(x, 1) \leq Mf(x)$. We may suppose that $x = 0$. But

Fig. 5.2 Approximation by
characteristic functions of
intervals

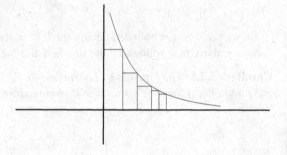

$$u(0, 1) = \int P_1(t) f(t)\, dt\,.$$

Of course

$$P_1(t) = \frac{c_n}{(1 + t^2)^{(n+1)/2}}\,.$$

In polar coordinates, write

$$P_1(r) = \frac{c_n}{(1 + r^2)^{(n+1)/2}}\,.$$

This may be approximated by a sum of characteristic functions of intervals with
total weight 1. See Fig. 5.2. Thus

$$\int P_1(t) f(t)\, dt \leq \sum_j w_j M f(0) \leq M f(0)\,.$$

By a familiar device, once we notice that for $|t| < \alpha y$ we have $P_y(x - 1) \leq C_\alpha P_y(x)$, we see that **(2)** follows from **(1)**.

For **(3)**, we write

$$u(x, y) - f(x_0) = PI(f(x) - f(x_0))$$

and also write

$$f(x) - f(x_0) = \chi_B (f(x) - f(x_0)) + \chi_{c_B}(f(x) - f(x_0)) \equiv T_1 + T_2\,.$$

Now $P_y(T_2)(x) \to 0$ as $(x, y) \to (x_0, 0)$ by localization. Also $PI(T_1)$ is
dominated by the maximal function of T_1. Complete the proof now by a standard
argument.   □

Theorem 5.2.10 (CAL) *Suppose that u is a harmonic function on $\mathbb{R}^{n+1}_+$. Let $E \subseteq \mathbb{R}^n$ have positive measure and suppose that u is nontangentially bounded at almost every $x_0 \in E$. Then u has nontangential limits at almost every point of E.*

Proof By a uniformization, which we confirm later, we may suppose that, for one value of $\alpha > 0$, $h > 0$, we have that u is bounded on $\mathcal{R} = \cup_{x \in E} \Gamma^h_\alpha(x)$ by 1. We also suppose that E is compact.

We will show that u has nontangential limits almost everywhere on E if the limit is taken only through cones of aperture α. This is clearly a weaker conclusion than that of the theorem, but we will worry about this later. Let

$$f_k(x) = \begin{cases} u(x, 1/k) & \text{for } (x, 1/k) \in \mathcal{R}, \\ 0 & \text{for } (x, 1/k) \notin \mathcal{R}. \end{cases}$$

Let $U_k(x, y) = P_y * f_k(x)$. The f_k form a bounded subset of $L^\infty(\mathbb{R})$, so there is a weak-$*$ convergent subsequence $f_{k_j} \to f$ in L^∞. Let $U = PI(f)$.
   Write

$$u(x, y + 1/k) = U_k(x, y) + V_k(x, y)$$

and

$$u(x, y) = U(x, y) + V(x, y).$$

We claim that

$$V_{k_j}(x, y) \to V(x, y)$$

pointwise as $j \to \infty$. This is clear since $U_{k_j} \to U$ pointwise and $u(x, y+1/k_j) \to u(x, y)$ pointwise.

We know that U has nontangential limit f almost everywhere. We will show that V has nontangential limit 0 almost everywhere. From this it will follow that u has nontangential limit f.

In order to see that V has nontangential limit 0, we shall show that it is dominated by a function that clearly has nontangential limit 0. Since $V_k \to V$, it suffices to show that V_k has limit 0 on E. Let χ be the characteristic function of ^cE. Then $PI(\chi)$ is harmonic and bounded by 1. We show that, for the right choice of C,

$$C \cdot (PI(\chi) + y) \geq |V_k(x, y)|. \tag{$\star$}$$

Clearly the function on the left has boundary limit 0 almost everywhere on E.

Since both function in ($\star$) are harmonic, it suffices to check the inequality ($\star$) on the boundary of $\mathcal{R}$. Now $|V_k| \leq 2$ on $\partial \mathcal{R}$ so, by choosing $C \geq 2/h$, we can guarantee that ($\star$) holds on $\partial \mathcal{R} \cap \{y = h\}$. Now if $x_0 \in E$, then $(x_0, 1/k) \in$

Fig. 5.3 The case $0 < y < h$

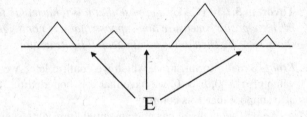

$\{(x, 1/k) \in \mathcal{R}\}$ so $V_k = 0$ at $(x_0, 0)$. Thus ($\star$) is trivial on $\partial \mathcal{R} \cap \{y = 0\}$. It remains only to check ($\star$) when $0 < y < h$. See Fig. 5.3.

But if $y > 0$ we have, for $\mathcal{U} = PI\mathcal{F}, \mathcal{F} \geq 0$, that

$$\mathcal{U}(x, y) = \int P_y(t)\mathcal{F}(x - t)\, dt \geq \int_{|t| \leq \alpha y} P_y(t)\mathcal{F}(x - t)\, dt$$

and

$$P_y(t) = \frac{c_n y}{(y^2 + |t|^2)^{(n+1)/2}}$$

$$\geq \frac{c_n y}{(y^2 + \alpha^2 y^2)^{(n+1)/2}}$$

$$\geq C_{n,\alpha} y^{-n}$$

$$\geq C'_{n,\alpha} (\Omega_n(\alpha y))^{-1}$$

for $|t| \leq \alpha y$.

So

$$\mathcal{U}(x, y) \geq C'_\alpha (\Omega_n(\alpha y))^{-1} \int_{|t| \leq \alpha y} f(x - t)\, dt\,.$$

Also

$$PI(\chi_{c_E}) \geq C'_\alpha\,.$$

We now conclude ($\star$) provided that $C \geq C'_\alpha/2$. Now we can use a density argument to see why the above uniformization is OK. $\qquad\square$

Lemma 5.2.11 *Suppose that* $E \subseteq \mathbb{R}^n$ *is closed. Let* $d(x, E)$ *denote the distance from* x *to* E. *Then, for almost every* $x \in E$, $d(x + y, E) = o(|y|)$.

Remark This refines the trivial fact that $d(x + y, E) = O(|y|)$.

Proof of the Lemma Let x be a point of density of E. Pick $\epsilon > 0$. Suppose that there is no point of E in $B(x+y, \epsilon|y|)$ for y arbitrarily near to 0. Then, for arbitrarily small y,

$$\frac{m(B(x, (1+\epsilon)|y|) \cap E)}{m(B(x, (1+\epsilon)|y|))} \leq \frac{[(1+\epsilon)^n|y|^n - \epsilon^n|y|^n]\Omega_n}{(1+\epsilon)^n|y|^n\Omega_n}$$

$$= \frac{(1+\epsilon)^n - \epsilon^n}{(1+\epsilon)^n}$$

$$< 1.$$

This contradicts the assumption that x was a point of density. Since almost every point of E is a point of density, we are done.   □

Now we apply the lemma to $\mathcal{R}$. We may write

$$\mathcal{R} = \{(x, y) : d(x, E) < \alpha y, 0 < y < h\}.$$

Thus

$$\partial\mathcal{R} \cap \{0 < y < h\} = \{(x, y) : y = d(x, E)/\alpha, 0 < y < h\}.$$

Let x_0 be a point of density of E and choose any $\beta \in \mathbb{R}$, $\beta > \alpha$. We will show that $\Gamma_\beta^k(x_0) \subseteq \mathcal{R}$ for k sufficiently small. This will show that u has a nontangential (unambiguous) boundary value at x_0 since we know that the limit is good when it is taken within the region $\mathcal{R}$.

Now $(x, y) \in \Gamma_\beta^k(x_0)$ means that $|x - x_0| < \beta y, 0 < y < k$. We need to see that, for such (x, y), with k small enough (depending on x_0) we have $d(x, E) < \alpha y$. But, by the lemma, there exists a $\delta > 0$ such that $|x - x_0| < \delta$ implies that

$$d(x_0 + (x - x_0), E) \leq \frac{\alpha}{2\beta} \cdot |x - x_0|.$$

It suffices to let $k = \delta/\beta$. Then $(x, y) \in \Gamma_\beta^k(x_0)$ implies that $|x - x_0| < \beta k = \delta$, hence

$$d(x, E) \leq \frac{\alpha}{2\beta} \cdot |x - x_0| < \alpha y$$

as desired.

For the uniformization, merely write E as a countable union of sets on which the heights and apertures of the cones of boundedness can be taken to be constant, say $E = \cup_{j=1}^\infty E_j$. Pick $\eta > 0$. By using finitely many E_j, we may find a set $E_0 \subseteq E$ such that $|E \setminus E_0| < \eta/4$ and the same cone may be used at each point of E_0.

Finally choose $E_* \subseteq E_0$, E_* compact, with $|E \setminus E_*| < \eta/2$. We get nontangential limits as desired at almost every point of E_*. Since $\eta > 0$ was arbitrary, we are finished with Calderón's theorem.   □

PROBLEMS ASSOCIATED WITH THE LOCAL FATOU THEOREM

(1)   Suppose that $\mathcal{D} \subseteq \mathbb{R}^n$ is bounded, open, with smooth (C^∞) boundary. Then essentially the same sort of local Fatou theorem holds with about the same proof. There are classical estimates for the Green's function on such a domain $\mathcal{D}$ which enable one to estimate the Poisson kernel. The problem is, how much smoothness does $\partial\mathcal{D}$ require? In fact $C^{1+\epsilon}$ boundary will do (see [WID]). For a domain with only Lipschitz boundary, the result holds almost everywhere with respect to harmonic measure. That is to say, the harmonic measure is that given by the Poisson kernel. It is not clear that this measure is equivalent to Lebesgue measure (although that equivalence was later proved by Dahlberg [DAH]). One can actually formulate the problem on any domain where almost every boundary point has a tangent plane.

HARMONICITY AND SUBHARMONICITY IN $\mathbb{R}^n$

Recall that if u is harmonic on $\mathcal{D} \subseteq \mathbb{R}^2$, $u > 0$, then

$$\Delta(u^p) = p(p-1)u^{p-2}|\nabla u|^2 .$$

Proof We calculate that

$$\Delta(u^p) = 4\frac{\partial}{\partial z}\frac{\partial}{\partial \overline{z}}(u^p)$$

$$= \frac{\partial}{\partial z}\left(4pu^{p-1}\frac{\partial u}{\partial \overline{z}}\right)$$

$$= 4p(p-1)u^{p-2}\frac{\partial u}{\partial \overline{z}}\frac{\partial u}{\partial z} + 4pu^{p-1}\frac{\partial^2 u}{\partial z \partial \overline{z}}$$

$$= p(p-1)u^{p-2}\cdot 4\cdot\frac{\partial u}{\partial \overline{z}}\frac{\partial u}{\partial z}$$

$$= p(p-1)u^{p-2}|\nabla u|^2 .$$

That completes the proof.   □

Also, if F is analytic on $\mathcal{D} \subseteq \mathbb{R}^2$, then

$$\Delta(|F|^p) = p^2|F|^{p-2}|F'|^2.$$

This is true whether or not F has zeros, but here we suppose that $F \neq 0$.

Proof We calculate that

$$\Delta(|F|^p) = 4\frac{\partial}{\partial z}\frac{\partial}{\partial \overline{z}}(F^{p/2}\cdot\overline{F}^{p/2})$$

$$= 4\cdot\frac{p}{2}\cdot F^{(p/2)-1}\cdot F'\cdot\frac{p}{2}\cdot\overline{F}^{(p/2)-1}\cdot\overline{F'}$$

$$= p^2|F|^{p-2}\cdot|F'|^2.$$

That gives the result. □

It follows that, if u is harmonic, then $|u|^p$ is subharmonic for $p \geq 1$. If F is holomorphic, then $|F|^p$ is subharmonic for $p > 0$. Moreover, $\log |F|$ is subharmonic. This result is on $\mathbb{R}^2$ only.

5.3 Cauchy–Riemann Systems

Now we consider $\mathbb{R}^n$. Observe that we can think of a holomorphic function on $\mathbb{R}^2$ as a pair of harmonic functions $F = u + iv$ such that $u_x = v_y$ and $u_y + v_x = 0$. (these are just the Cauchy–Riemann equations.) Therefore, on $\mathbb{R}^n$, we define a *Cauchy–Riemann system* to be an n-tuple of functions $(u_1, u_2, \ldots, u_n)$ which satisfy the conditions of M. Riesz:

(i) $\sum \dfrac{\partial u_j}{\partial x_j} = 0.$

(ii) $\dfrac{\partial u_j}{\partial x_k} = \dfrac{\partial u_k}{\partial x_j}.$

Claim If $(u_1, u_2, \ldots, u_n)$ is a Cauchy–Riemann system, then each u_j is harmonic. In fact, the system is the gradient of a harmonic function. To see this, set $w = \sum u_j \, dx_j$. Then condition (i) says that $\delta w = 0$ (where δ is the operator in exterior algebra which is adjoint to d) and (ii) says that $dw = 0$. These are the Hodge equations. So we may define, for x_0 fixed,

$$H(x) = \int_{x_0}^{x} w,$$

and the definition is, by Stokes's theorem, independent of the choice of path. Moreover, it is easily seen that $\triangle H = 0$. Conversely, if H is harmonic on $\mathbb{R}^n$, then

$$\nabla H = \left(\frac{\partial H}{\partial x_j} \right) = (u_j)$$

is a Cauchy–Riemann system. We generally assume that the u_j are real-valued.

Lemma 5.3.1 *If $(u_1, u_2, \ldots, u_n)$ is a Cauchy–Riemann system with $|u|^2 \equiv \sum_{j=1}^{n} |u_j|^2$, then*

$$\triangle(|u|^p) \geq C_p |u|^{p-2} |\nabla u|^2$$

near a point where $|u| > 0$. Here $C_p \geq 0$ and $p \geq (n-2)/(n-1)$. One can show that $C_p > 0$ if $p > (n-2)/(n-1)$.

Proof Now

$$\frac{\partial}{\partial x_j}(|u|^p) = \frac{\partial}{\partial x_j}\left[(u, u)^{p/2}\right]$$

$$= \frac{p}{2}(u, u)^{(p/2)-1} \cdot 2 \cdot (u_{x_j}, u)$$

$$= p|u|^{p-2} \cdot (u_{x_j}, u).$$

Thus

$$\frac{\partial^2}{\partial x_j^2}|u|^p = p(p-2)|u|^{p-4} \cdot (u_{x_j}, u)^2 + p|u|^{p-2}\left[(u_{x_j x_j}, u) + (u_{x_j}, u_{x_j})\right].$$

Summing over j, we get

$$\Delta(|u|^p) = p(p-2)|u|^{p-4}\sum_{j=1}^{n}(u_{x_j}, u)^2 + p|u|^{p-2}\sum_{j=1}^{n}|u_{x_j}|^2$$

$$= p|u|^{p-4}\left\{(p-2)\sum_{j=1}^{n}(u_{x_j}u)^2 + |u|^2\sum_{j=1}^{n}|u_{x_j}|^2\right\}. \tag{*}$$

Recall that, when $n = 2$, $\Delta|F|^p = p^2|F|^{p-2}|F'|^2$ and we seek analogues in the present situation. Now, when $p \geq 2$, it is clear that $(*) \geq 0$. When $p < 2$, the first term in $(*)$ is ≤ 0. As an interesting example, when $p = 1$,

$$(p-2)\sum_{j=1}^{n}(u_{x_j}, u)^2 + |u|^2\sum_{j=1}^{n}|u_{x_j}|^2 \geq 0$$

just because

$$\sum_{j=1}^{n}(u_{x_j}, u)^2 \leq |u|^2\sum_{j=1}^{n}|u_{x_j}|^2$$

by Schwarz's inequality.

More generally, for $p < 2$, we need to see that

$$\sum_{j=1}^{n}(u_{x_j}, u)^2 \leq \frac{1}{2-p}|u|^2\sum_{j=1}^{n}|u_{x_j}|^2.$$

This reduces to a problem about matrices. Consider the Hession matrix

$$M(u) \equiv \left(u_{jk}\right) = \left(\frac{\partial u_j}{\partial x_k}\right) = \left(\frac{\partial u_k}{\partial x_j}\right).$$

This is a symmetric matrix with trace 0. We need to see that

$$|M(u)|^2 \leq \frac{1}{2-p}\|M\|_{\mathrm{HS}}^2 \cdot |u|^2.$$

Here $u = (u_1, u_2, \ldots, u_n)$ and $\|\ \|_{\mathrm{HS}}$ is the Hilbert–Schmidt norm. Equivalently, it suffices to see that

$$\|M\|^2 \leq \frac{1}{2-p}\|M\|_{\mathrm{HS}}.$$

It is well known that

$$\|M\|^2 \leq \|M\|_{\mathrm{HS}}^2,$$

so the case $p \geq 1$ is clear.

So we need to treat the case $p < 1$. It suffices, by invariance, to consider only diagonal matrices with real entries. So we need to know that

$$\sup_j |\lambda_j|^2 \leq \frac{1}{2-p}\sum_j \lambda_j^2$$

whenever $\sum_j \lambda_j = 0$, $\lambda_j \in \mathbb{R}$. Thus it is enough to see that

$$|\lambda_k|^2 \leq \frac{1}{2-p}\sum_j \lambda_j^2$$

for each k. We may suppose without loss of generality that $k = 1$. Now

$$\lambda_1 = -(\lambda_2 + \cdots + \lambda_n)$$

so

$$\lambda_1^2 \leq \left(\sum_{j=2}^n |\lambda_j|\right)^2 \leq (n-1)(\lambda_2^2 + \cdots \lambda_n^2)$$

by Cauchy–Schwarz–Bunjakovski.

That is to say,

$$n\lambda_1^2 \leq (n-1)\sum_{j=1}^{n}\lambda_j^2$$

or

$$\lambda_1^2 \leq \frac{n-1}{n}\sum_{j=1}^{n}\lambda_j^2.$$

So, provided that $1/(2-p) \geq (n-1)/n$, we have our result. But this just says that $p \geq (n-2)/(n-1)$, as desired.   □

Corollary 5.3.2 *If u is a Cauchy–Riemann system on $\mathcal{D} \subseteq \mathbb{R}^n$, then $|u|^p$ is subharmonic for $p > 0$. Here*

$$|u|^p \equiv \left(\sum_{j=1}^{n}|u_j|^2\right)^{p/2}$$

when $u = (u_1, u_2, \ldots, u_n)$.

Now recall that

$$P_y(x) = \frac{c_n y}{(|x|^2 + y^2)^{(n+1)/2}}$$

and

$$\int P_y(x)e^{2\pi i x \cdot \xi}\, dx = e^{-2\pi|\xi|y}$$

for $y > 0$. We now define n conjugate Poisson kernels by

$$\int Q_y^j(x)e^{2\pi i x \cdot \xi}\, dx = i e^{-2\pi|\xi|y} \cdot \frac{\xi_j}{|\xi|}. \tag{$\star_1$}$$

One can compute that

$$Q_y^j(x) = \frac{c_n x_j}{(|x|^2 + y^2)^{(n+1)/2}}. \tag{$\star_2$}$$

It is pedestrian to check that

$$(P_y(x), Q_y^1(x), \ldots, Q_y^n(x))$$

is a Cauchy–Riemann system on $\mathbb{R}_+^{n+1}$. In fact this system is the gradient of

$$H(x, y) = \frac{C}{(|x|^2 + y^2)^{(n-1)/2}},$$

which is harmonic. For $f \in L^p(\mathbb{R}^n)$, define

$$u_0(x, y) = P_y * f(x)$$
$$u_k(x, y) = Q_y^k * f(x) \text{ for } k = 1, \ldots, n \text{ and } y > 0.$$

We claim that $(u_0, u_1, \ldots, u_n)$ is a Riesz system on $\mathbb{R}_+^{n+1}$. This is in fact clear from the above. We will see below that

$$\sup_{y>0} \int_{\mathbb{R}^n} |u_j(x, y)|^p \, dx < \infty,$$

hence each u_j has a boundary function in L^p, provided that $1 < p < \infty$. Call the boundary function f_j. We would like to be able to obtain the f_js directly from f. We might hope that this can be accomplished by convolving f with the kernel obtained by setting $y = 0$ in $Q_y^j(x)$.

Thus let

$$Q_0^k(x) \equiv \frac{c_n x_k}{|x|^{n+1}} = \frac{\Omega(x)}{|x|^n},$$

where $\Omega(x)$ is homogeneous of degree 0, is smooth, and is odd. We remark now that $(\star_1)$ and $(\star_2)$ are consistent because

$$\frac{\partial}{\partial \xi_j} e^{-2\pi |\xi| y} = -2\pi y e^{-2\pi |\xi| y} \frac{\xi_j}{|\xi|}.$$

For taking inverse Fourier transforms of this expression yields

$$x_j P_y(x) = y Q_y^j(x)$$

or

$$Q_y^j(x) = \frac{c_n x_j}{(|x|^2 + y^2)^{(n+1)/2}}.$$

Now we shall define, on $L^2(\mathbb{R}^n)$,

$$\widehat{R_j f}(\xi) = \frac{i \xi_j}{|\xi|} \cdot \widehat{f}(\xi).$$

By Plancherel's theorem,

$$\|R_j\|_2 \le 1$$

and

$$\sum_{j=1}^{n} R_j^2 = -I .$$

Also, for $f \in L^2$,

$$Q_y^j * f \to R^j f$$

in the L^2 topology as $y \to 0$. Now we define

$$H^2(\mathbb{R}_+^{n+1}) = \{(u_0, u_1, \ldots, u_n) : \text{ the } (n+1)\text{-tuple is a}$$

$$\text{C-R system and } \sup_{y>0} \int_{\mathbb{R}^n} |u(x, y)|^2 \, dx \equiv \|u\|_{H^2}^2 < \infty\} .$$

MAIN PROPOSITION *The function* $u \in H^2(\mathbb{R}_+^{n+1})$ *if and only if there exists an* $f \in L^2(\mathbb{R}^n)$ *such that*

$$u_0(x) = P_y * f(x)$$
$$u_j(x) = Q_y^j * f(x) = P_y * (R^j f)(x) , \quad j = 1, \ldots, n$$

and

$$\|u\|_{H^2}^2 = 2\|f\|_2^2 .$$

Proof Given $f \in L^2(\mathbb{R}^n)$, then by definition of the u_j,

$$\int |u_0(x, y)|^2 \, dx \le \|f\|_2^2$$

and

$$\int |u_j(x, y)|^2 \, dx \le \|R^j f\|_2^2 \le \|f\|_2^2 .$$

Moreover,

$$\sum_j \|R_j f\|_2^2 = \|f\|_2^2$$

and

$$\sup_{y>0} \int |u_0(x, y)|^2 \, dx = \sup_{y>0} \int |P_y * f|^2 \, dx = \|f\|_2^2,$$

$$\sup_{y>0} \int |u_j(x, y)|^2 \, dx = \sup_{y>0} \int |P_y * R^j f|^2 \, dx = \|R^j f\|_2^2.$$

Thus

$$\|u\|_{H^2}^2 = 2\|f\|_2^2.$$

Conversely, if u is given, each u_j harmonic and satisfying

$$\sup_{y>0} \int |u_j(x, y)|^2 \, dx < \infty,$$

then each u_j has a boundary function $f_j \in L^2$ and $u_j = P_y * f_j$. Now

$$u_0(x, y) = \left(e^{-2\pi |\xi| y} \widehat{f_0}(\xi) \right)^{\vee}$$

$$u_j(x, y) = \left(e^{-2\pi |\xi| y} \widehat{f_j}(\xi) \right)^{\vee}.$$

Since

$$\frac{\partial}{\partial y} u_j = \frac{\partial}{\partial x_j} u_0,$$

it follows that

$$|\xi| \widehat{u_j} = i\xi_j \widehat{u_0},$$

hence

$$\widehat{f_j} = \frac{i\xi_j}{|\xi|} \widehat{f_0},$$

and finally

$$u_j(x) = Q_y^j * f_0(x).$$

By the first part, $\|u\|_{H^2}^2 = 2\|f\|_{L^2}^2$. $\qquad\qquad\square$

Now we have defined, for $f \in L^2$,

$$R_j f = \left(\frac{i\xi_j}{|\xi|} \cdot \widehat{f}(\xi) \right)^{\vee} .$$

But we would like to see if one can use the kernel $c_n x_j / |x|^{n+1}$ obtained by letting $y = 0$ in the kernel for Q_y^j. Let us define

$$R_j^{\epsilon}(f)(x) = \int_{|y| \geq \epsilon} f(x - y) \frac{y_j}{|y|^{n+1}} \, dy .$$

By Schwarz's inequality, $R_j^{\epsilon}(f)$ is a measurable, finite-almost-everywhere, function for $f \in L^p$, $1 \leq p < \infty$.

Lemma 5.3.3 *For $f \in L^p(\mathbb{R}^n)$, $1 \leq p < \infty$, we have $R_j^{\epsilon} f - Q_{\epsilon}^j f \to 0$ in L^p and almost everywhere as $\epsilon \to 0$.*

The reader can check, by a computation that we have done before on the circle, that the kernel for $R_j^{\epsilon} - Q_{\epsilon}^j$ is in L^1 so that the expression $R_j^{\epsilon} f - Q_{\epsilon}^j f$ makes sense for $f \in L^{\infty}$. However, the terms do not make sense individually.

Of course, $R_j^{\epsilon} f = K_{\epsilon}^j * f$, where

$$K_{\epsilon}^j(x) = \begin{cases} \frac{c_n x_j}{|x|^{n+1}} & \text{for} \quad |x| \geq \epsilon \\ 0 & \text{for} \quad |x| < \epsilon . \end{cases}$$

Moreover,

$$K_j^{\epsilon} - Q_{\epsilon}^j = \epsilon^{-n} \Phi^j(x/\epsilon) ,$$

where, we claim,

(1) $|\Phi^j(x)| \leq A(1 + |x|)^{-n-1}$.
(2) $\int \Phi^j(x) \, dx = 0$.

To see this claim, notice that

$$\Phi^j(x) = \begin{cases} \frac{-c_n x_j}{(|x|^2+1)^{(n+1)/2}} & \text{if} \quad |x| \leq 1 \\ \frac{c_n x_j}{|x|^{n+1}} - \frac{c_n x_j}{(|x|^2+1)^{(n+1)/2}} & \text{if} \quad |x| > 1 . \end{cases}$$

Clearly Φ^j is odd, so **(2)** is clear. Moreover, as $x \to \infty$,

$$|\Phi^j(x)| = \left| \frac{c_n x_j}{|x|^{n+1}} - \frac{c_n x_j}{(|x|^2 + 1)^{(n+1)/2}} \right|$$

$$\leq C|x| \left| \frac{(|x|^2 + 1)^{(n+1)/2} - |x|^{n+1}}{|x|^{n+1}(|x|^2 + 1)^{(n+1)/2}} \right|$$

$$\leq \frac{C_n |x| \cdot |x|^{n-1}}{|x|^{2n+1}}$$

$$\leq \frac{C}{|x|^{n+1}}.$$

Theorem 5.3.4 *Suppose that Ω is homogenous of degree 0 on $\mathbb{R}^n$, that is to say, $\Omega(\delta x) = \Omega(x)$ for all $\delta > 0$. Suppose further that $\Omega|_{S^{n-1}} \in L^1(S^{n-1})$, and also that Ω is odd (i.e., $\Omega(-x) = -\Omega(x)$). Let*

$$T_{\epsilon, N} f(x) = \int_{\epsilon \leq |y| \leq N} \frac{\Omega(y)}{|y|^n} f(x - y) \, dy$$

for $f \in L^p$, $1 \leq p \leq \infty$. Then

$$\|T_{\epsilon, N} f\|_{L^p} \leq A_p \|f\|_{L^p} \,, \quad 1 < p < \infty, \text{ with } A_p \text{ independent of } \epsilon \text{ and } N.$$

Proof We first do the case $n = 1$. In this case the only possible Ω is a constant multiple of $\operatorname{sgn} x$. So we let $\Omega(x) = \operatorname{sgn} x$ and

$$T_{\epsilon, N}(f)(x) = (T_\epsilon f - T_N f)(x),$$

where

$$T_a f(x) \equiv \int_{|y| \geq a} \frac{f(x - y)}{y} \, dy \,, \quad a > 0.$$

We know that

$$\|T_a f\|_p \leq A_p \|f\|_p \,, \quad 1 < p < \infty$$

by the Theorem of M. Riesz which we have proved. So the $T_{\epsilon, N}$ are bounded in norm by $2A_p$.

The Case of Arbitrary Dimension n

Define

$$H_{\epsilon, N} f(x) = \int_{N \geq |y| \geq \epsilon} \frac{f(x - y)}{y} \, dy.$$

If $\eta \in S^{n-1} \subseteq \mathbb{R}^n$ is a unit vector, set

$$H_{\epsilon,N}^{\eta}(f)(x) = \int_{\epsilon \leq t \leq N} \frac{f(x - t\eta)}{t} \, dt + \int_{-N \leq t \leq -\epsilon} \frac{f(x - t\eta)}{t} \, dt \, .$$

Here t is a real variable.

We claim that

$$\|H_{\epsilon,N}^{\eta}\|_p \leq A_p$$

with A_p independent of ϵ, N, η. This assertion follows from the one-dimensional result. Now

$$H_{\epsilon,N}(f)(x) = \int_{\epsilon \leq |y| \leq N} \frac{\Omega(y)}{|y|^n} f(x - y) \, dy$$

$$= \int_{S^{n-1}} \int_{\epsilon \leq r \leq N} \frac{\Omega(\eta r) f(x - \eta r)}{r^n} \, dr \, d\eta \, .$$

Here $y = \eta \cdot r$ and $r = |y| > 0$. This last

$$= \iint_{\epsilon \leq r \leq N} \frac{\Omega(\eta r) f(x - \eta r)}{r} \, dr \, d\eta$$

$$= \frac{1}{2} \int_{\substack{\epsilon \leq t \leq N \\ -N \leq t \leq -\epsilon}} \frac{\Omega(\eta t) f(x - \eta t)}{t} \, dt \, d\eta$$

by oddness. But then we can say that this equals

$$\frac{1}{2} \int_S H_{\epsilon,N}^{\eta}(f)(x)\Omega(\eta) \, d\eta \, .$$

Thus

$$\|T_{\epsilon,N} f\|_{L^p} \leq \frac{A_p}{2} \|\Omega\|_{L^1(S^{n-1})} \cdot \|f\|_{L^p} \, .$$

That ends the argument for general n. □

Corollary 5.3.5 *Set*

$$R_j^{\epsilon} f(x) = c_n \int_{|y| \geq \epsilon} f(x - y) \frac{y_j}{|y|^{n+1}} \, dy$$

for $f \in L^p$, $1 < p < \infty$. Then

(1) $\|R_j^{\epsilon} f\|_p \leq A_p \|f\|_p$, $1 < p < \infty$.

(2) $R_j^\epsilon f \overset{\text{def}}{\to} R_j f$ in L^p norm as $\epsilon \to 0$.

(3) $\|R_j f\|_p \le A_p \|f\|_p$, where $R_j f$ is defined by the limit in (2).

Proof By (1), which is clear, we need only check (2) on a dense subset of L^p, say C_c^1.

Let $\epsilon_2 > \epsilon_1 > 0$ and $f \in C_c^1$. Then

$$R_j^{\epsilon_1} f(x) - R_j^{\epsilon_2} f(x) = c_n \int_{\epsilon_1 \le |y| \le \epsilon_2} \frac{f(x-y)y_j}{|y|^{n+1}} dy$$

$$= c_n \int_{\epsilon_1 \le |y| \le \epsilon_2} [f(x-y) - f(x)] \cdot \frac{y_j}{|y|^{n+1}} dy \qquad (*)$$

since

$$\int_{\epsilon_1 \le |y| \le \epsilon_2} \frac{y_j}{|y|^{n+1}} dy = 0 .$$

Now $|f(x-y) - f(x)| \le A|y|$, with A depending on f, since $f \in C_c^1$. Thus

$$(*) \le c_n \int_{\epsilon_1 \le |y| \le \epsilon_2} \frac{dy}{|y|^{n-1}} \to 0$$

as $\epsilon_1, \epsilon_2 \to 0$. This shows that the R_j^ϵ are pointwise Cauchy. By Lebesgue dominated convergence, we find that they are L^p Cauchy. If we define $R_j f$ to be the resulting limit, then (3) is clear.

Remark 5.3.6 We may now prove an analogue of the theorem above for *even* kernels. Normally with the same hypotheses except that Ω even. Write

$$T_{\epsilon,N} = -(-I T_{\epsilon,N}) = -\sum_j R_j(R_j T_{\epsilon,N}) .$$

Then, for each j, $R_j T_{\epsilon,N}$ has an odd kernel so that we may apply the theorem. Since each R_j is a bounded operator, we are done.

Observe that this theory works because our operators are all convolution operators (hence they commute with translations) which commute with dilations.

5.4 A Characterization of H^p, $p > 1$

Definition 5.4.1 For $1 < p < \infty$, $(u_0, u_1, \ldots, u_n) = F(x, y)$, $(x, y) \in \mathbb{R}_+^{n+1}$, we set

$$H^p(\mathbb{R}_+^{n+1}) = \{F : \sup_{y>0} \int_{\mathbb{R}^n} |F(x, y)|^p dx < \infty\}.$$

Theorem 5.4.2 *The function $F \in H^p$ if and only if there exists an $f \in L^p(\mathbb{R}^n)$ such that*

$$u_0(x, y) = PIf$$

and

$$u_j(x, y) = PI(R_j f) = Q_y^j * f \ , \ j = 1, \dots, n.$$

Also $\|F\|_{H^p} \approx \|f\|_{L^p}$.

Proof The reverse direction is clear by the boundedness of R_j on L^p, $1 < p < \infty$.

For the forward direction, if $F \in H^p$, then all components are uniformly bounded in L^p, $1 < p$, so each is the Poisson integral of an L^p function. Write

$$PI(f_j) = u_j(x, y) \ , \ j = 0, 1, \dots, n.$$

We wish to see that $f_j = R_j(f_0)$.

Let $v_j = Q_y^j(f_0)$. Thus we obtain two Cauchy–Riemann systems:

$$\begin{pmatrix} u_0 \\ v_1 \\ \vdots \\ v_n \end{pmatrix} \qquad \text{and} \qquad \begin{pmatrix} u_0 \\ u_1 \\ \vdots \\ u_n \end{pmatrix}.$$

By subtraction, we get a Cauchy–Riemann system $U_j = u_j - v_j$ with first entry 0. We want to see that the whole system is 0. Now

$$\sup_{y>0} \int_{\mathbb{R}^n} |U_j(x, y)|^p \, dx < \infty$$

and

$$\frac{\partial U_j}{\partial y} = \frac{\partial U_0}{\partial x_j} \equiv 0.$$

Thus each U_j is a harmonic function of x only and in L^p in the x variable. So each U_j is identically 0 by the mean value property. (There are no L^p harmonic functions on $\mathbb{R}^n$ except 0.) $\square$

The Case $(n-1)/n < p < \infty$

Define $H^p(\mathbb{R}_+^{n+1})$, $p > (n-1)/n$, as before.

Theorem 5.4.3 *Suppose* $F \in H^p$, $p \leq 1$. *Then* $\lim_{y \to 0} F(x, y)$ *exists for almost every* $x \in \mathbb{R}^n$ *and, more generally, the limit exists almost everywhere nontangentially.*

  Moreover,

$$\int_{\mathbb{R}^n} |F(x, y) - F(x)|^p \, dx \to 0 \text{ as } y \to 0. \qquad (*)$$

If we let

$$\Gamma_\alpha(x_0) = \{(x, y) : |x - x_0| < \alpha y, 0 < y < \infty\},$$

then

$$\left(\int_{\mathbb{R}^n} \sup_{(x,y) \in \Gamma_\alpha(x_0)} |F(x, y)|^p \, dx_0 \right)^{1/p} \leq A_{p,\alpha} \|F\|_{H^p} .$$

  All of the above, except $(*)$, *persists when* $p = \infty$.

Proof (Outline) Let $p_0 = (n - 1)/n$. Consider

$$|F|^{p_0} = \left(\sum_{j=0}^{n} |u_j|^2 \right)^{p_0/2} \equiv S(x, y).$$

Now $S(x, y)$ is subharmonic and continuous. Consider $S(x, y + 1/k)$, which, for each $k > 0$, is subharmonic and continuous up to the boundary. Let $h_k(x, y) = PI(S(x, 1/k))$. Thus $h_k(x, y) \geq 0$. Set $q = p/p_0 > 1$. We have

$$\sup_{y>0} \int_{\mathbb{R}^n} |S(x, y)|^q \, dx < \infty.$$

In particular, $S(x, 1/k) \in L^q(\mathbb{R}^n)$, uniformly in k. By a standard argument, using the mean value property,

$$S(x, y) \leq cy^{-n/q} .$$

So, by a limiting argument, we get, using harmonic majorization, that

$$S(x, y + 1/k) \leq h_k(x, y)$$

on $\mathbb{R}^{n+1}_+$.

  Now we may, in the usual manner, find a subsequence k_j so that $h_{k_j}(x, y)$ converges weak-$*$ to a function h. It is not difficult to check that h is harmonic (just use the mean value property). By the boundedness assumption,

$$h(x, y) = PI(h)(x)$$

for $h \in L^q(\mathbb{R}^n)$. Thus $S(x, y) \le PI(h)$. Since h is the Poisson integral of a function in $L^q, q > 1$, we have that h is nontangentially bounded almost everywhere. Hence so is S and so is u. But by the local Fatou theorem, we see that u has nontangential limits almost everywhere pointwise. The other conclusions of the theorem follow by standard arguments at this point.   □

OPEN PROBLEM #1 What happens when $p \le (n-1)/n$? There is an H^p theory for $p \le (n-1)/n$ which is based on a stronger definition, so what is really asked here is whether the stronger definition and the desired one are equivalent.

Fact For general n, let $u(x, y)$ be a harmonic function with

$$\sup_{y>0} \int |u(x, y)|^p \, dx < \infty \; , \; p < 1.$$

Then the boundary values need not exist in any sense.

OPEN PROBLEM #2 The H^p theory for $1 < p < \infty$ on $\mathbb{R}^n$ is equivalent to the boundedness of the Riesz transforms which follows from (and implies) the boundedness of the Hilbert transform which is equivalent to the H^p theory, $1 < p < \infty$, on $\mathbb{R}^1$. Is the H^p theory on $\mathbb{R}^n$, $p \le 1$, reducible to the one-dimensional theory in any analogous sense?

Now let $B(\mathbb{R}^n)$ be the space of continuous linear functionals on $C_c(\mathbb{R}^n)$. We can identify $B(\mathbb{R}^n)$ with the space of finite signed measures. One can also imbed

$$L^1(\mathbb{R}^n) \hookrightarrow B(\mathbb{R}^n)$$
$$f(x) \mapsto f(x) \, dx .$$

If $d\mu \in B(\mathbb{R}^n)$, then we set $R_j(d\mu) = dv$, where

$$\widehat{v}(\xi) = \frac{i\xi}{|\xi|} \cdot \widehat{\mu}(\xi) .$$

We have the following corollaries of the theorems:

Corollary 5.4.4 *The function $F \in H^1$ if and only if there exists an $f \in L^1$ such that $R_j f \in L^1$, $j = 1, \ldots, n$, [This latter condition is equivalent with there exists $f_j \in L^1$ such that $R_j f = f_j$.] and so that $F = (u_0, u_1, \ldots, u_n)$, $u_0 = PI(f)$, $u_j = PI(f_j)$, and*

$$\|F\|_{H^1} \approx \|f\|_{L^1} + \sum_{j=1}^{n} \|R_j f\|_{L^1} .$$

Corollary 5.4.5 *Suppose* $d\mu, d\mu_1, \ldots, d\mu_n \in B(\mathbb{R}^n)$ *are given and suppose further that* $d\mu_j = R_j(d\mu), j = 1, \ldots, n.$ *Then* $d\mu, d\mu_1, \ldots, d\mu_n$ *are absolutely continuous.*

Proofs of the Corollaries First we prove the second corollary. By hypothesis, the $u_j(x, y) = PI(d\mu_j)$ form a Cauchy–Riemann system. But, by the theorem, the $u_j(x, y)$ converge to their boundary values in L^1 norm. But also the u_j converge weak-$*$ to $d\mu_j$. For this to be consistent, we must have that $u_j(x, y) \to d\mu_j \in L^1$ as $y \to 0$. With this information in hand, the first corollary is clear. $\qquad\square$

Remark 5.4.6 One of the chief facts here is that if H is harmonic on $\mathbb{R}^{n+1}_+$, then ∇H is a Cauchy–Riemann system and $|\nabla H|^p$ is subharmonic if $p > (n-1)/n$. A generalization of this fact is that $|\nabla^k H|^p$ is subharmonic if $p \geq (n-1)/(n+k-1)$. Thus we may set

$$\frac{\partial^k H}{\partial y^k} = u_0(x, y).$$

Then, for each multi-index $(\alpha_0, \alpha_1, \ldots, \alpha_n)$ with $\sum_j \alpha_j = k$, we let

$$u_\alpha(x, y) = \left(\frac{\partial}{\partial x}\right)^\alpha H$$

(here x_0 denotes y). The u_α serve as conjugates of u_0. Instead of merely the Riesz transforms, we use monomials in the Riesz transforms. There are several first-order systems of partial differential equations associated with this generalized Cauchy–Riemann system of harmonic functions.

5.5 The Area Integral

As usual, we let

$$\Gamma_\alpha^h(x_0) = \{(x, y) : |x - x_0| < \alpha y, 0 < y < h\}$$

and

$$\Gamma_\alpha(x_0) = \{(x, y) : |x - x_0| < \alpha y\}.$$

Suppose that u is a given locally integrable function on $\mathbb{R}^{n+1}_+$. We define

$$S(u)(x_0) = \left(\int_{\Gamma_\alpha^h(x_0)} |\nabla u|^2 y^{1-n} \, dx dy\right)^{1/2}.$$

Note that this is an appropriate expression for Fourier analysis since it is quadratic. We use Γ_α^h for local theorems and Γ_α for global theorems.

Theorem 5.5.1 (The Local Theorem) *Let u be harmonic on $\mathbb{R}^{n+1}_+$. Then the set where $S(u)(x) < \infty$ and the set where u has nontangential boundary values are equal almost everywhere.*

Theorem 5.5.2 (The Global L^2 Analogue) *Let $f \in L^2(\mathbb{R}^n)$, $u = PI(f)$. Then, modulo an additive constant,*

$$\int_{\mathbb{R}^n} S^2(u)(x)\, dx = c_\alpha \|f\|_2^2.$$

This result was first observed by Lusin in 1930 in case $n = 1$. It was proved classically by Marcinkiewicz, Zygmund, and Spencer. We call S the *area integral* since, on $\mathbb{R}^2$, if $F = u + iv$ is holomorphic with $|F'|^2 = |\nabla u|^2$, then

$$\int_\Gamma |F'|^2\, dx dy$$

is the area of the image under F of Γ.

Now we shall prove the global or "stripped down" version of the area integral result. We let

$$\chi(x_0, x, y) = \begin{cases} 1 & \text{if } |x - x_0| < \alpha y \\ 0 & \text{otherwise.} \end{cases}$$

Here α is fixed throughout.

Hence

$$\int_{\mathbb{R}^n} \iint_{\Gamma_\alpha(x_0)} |\nabla u(x, y)|^2 y^{1-n}\, dx dy dx_0 = \iiint_{\mathbb{R}^n \times \mathbb{R}^n \times \mathbb{R}^n} |\nabla u(x, y)|^2 \cdot \chi(x_0, x, y) y^{1-n}\, dx dy dx_0.$$

$$(*)$$

But, for each fixed x and y, $\int \chi(x_0, x, y)\, dx_0 = c(\alpha y)^n$. Hence

$$(*) = c_1 \iint_{\mathbb{R}^{n+1}_+} |\nabla u(x, y)|^2 y\, dx dy$$

by Fubini.

If we let $\widehat{f}$ denote the Fourier transform of f and recall that

$$u(x, y) = \int_{\mathbb{R}^n} \widehat{f}(\xi) e^{-2\pi |\xi| y} e^{-2\pi i x \cdot \xi}\, d\xi$$

and

$$\left(\frac{\partial u}{\partial y}, \frac{\partial u}{\partial x_1}, \dots \frac{\partial u}{\partial x_j}, \dots, \frac{\partial u}{\partial x_n}\right) = \int_{\mathbb{R}^n} (-2\pi |\xi|, -2\pi i \xi_j) \widehat{f}(\xi) e^{-2\pi |\xi| y} e^{-2\pi i x \cdot \xi} \, d\xi \,,$$

then we have

$$\int_0^\infty \left(\int_{\mathbb{R}^n} |\nabla u(x, y)|^2 \, dx\right) y \, dy$$

$$= \int_0^\infty \int_{\mathbb{R}^n} 8\pi^2 |\xi|^2 |\widehat{f}(\xi)|^2 e^{-4\pi |\xi| y} \, d\xi \, y \, dy \quad \text{by Plancherel}$$

$$= \int 8\pi^2 |\xi|^2 |\widehat{f}(\xi)|^2 \, d\xi \int_0^\infty e^{-4\pi |\xi| y} y \, dy$$

$$= \int 8\pi^2 |\widehat{f}(\xi)|^2 \frac{1}{16\pi^2} \, d\xi$$

$$= \frac{1}{2} \int |\widehat{f}(\xi)|^2 \, d\xi$$

$$= \frac{1}{2} \|f\|_2^2 \,.$$

Now, in order to motivate the local theory, we give an alternative proof using Green's theorem.

Theorem 5.5.3 (Green's theorem) *Let $G(x, y)$ be smooth on $\mathbb{R}_+^{n+1}$ and vanishing at ∞. (these concepts will be made clear momentarily.) Then*

$$\int_{\mathbb{R}_+^{n+1}} y \Delta G(x, y) \, dx dy = \int_{\mathbb{R}^n} G(x, 0) \, dx \,.$$

[Here, for "smooth," C^2 will suffice. For "vanishing at ∞," we need enough so that when we apply the ordinary Green's theorem on a large semi-ball and let the radius go to ∞, then the boundary term vanishes.]

Also recall that $\frac{1}{2}\Delta |u|^2 = |\nabla u|^2$ when u is harmonic.

Now

$$\iint_{\mathbb{R}_+^{n+1}} |\nabla u(x, y)|^2 y \, dx dy = \frac{1}{2} \int_{\mathbb{R}_+^{n+1}} \Delta |u|^2 y \, dx dy$$

$$= \frac{1}{2} \int_{\mathbb{R}^n} |f(x)|^2 \, dx \,.$$

But we showed in the first proof that

$$\int_{\mathbb{R}^n} |S(u)(x)|^2 \, dx = c_\alpha \iint_{\mathbb{R}^{n+1}_+} |\nabla u(x, y)|^2 y \, dx dy.$$

This completes the proof. One should check that u vanishes swiftly enough at ∞.

$\square$

We wish to make this second proof more precise in order to get the local result. In this discussion we will always assume that our functions are real.

FIRST PART: We show that, if u is nontangentially bounded for $x_0 \in E$, then $S(u)(x_0) < \infty$. (this holds almost everywhere for any set E of positive measure.)
SECOND PART: Converse.

Proof of the First Part Let α, β, h, k be fixed, positive numbers with $\beta > \alpha$ and $k > h$. Assume that E is compact and that $|u(x, y)| \leq 1$ for all $(x, y) \in \cup_{x_0 \in E} \Gamma^k_\beta(x_0)$. We will show that $S(u)(x_0) < \infty$ for almost every $x_0 \in E$ with cones Γ^h_α. Ultimately, we make everything independent of the cone.

Lemma 5.5.4 *Suppose that u is harmonic on Γ^k_β with $|u| \leq 1$. Then $y|\nabla u| \leq C_{\alpha,\beta,h,k}$ on Γ^h_α.*

Proof There exists a $C > 0$ such that, for any point $(x, y) \in \Gamma^h_\alpha$, we have $B((x, y), cy) \subseteq \Gamma^k_\beta$. Let $\psi \in C^\infty_c(\mathbb{R}^{n+1})$ be radial with $\int_{\mathbb{R}^{n+1}} \psi \, dx dy = 1$. Write

$$\psi_\epsilon(x, y) = \epsilon^{-n-1} \psi(x/\epsilon, y/\epsilon).$$

We may suppose that ψ is supported in the unit ball.

Let u be bounded and harmonic in a subregion of $\mathbb{R}^{n+1}_+$. Let $u \equiv 0$ outside this region. Then $u(x, y) = u * \psi_\epsilon(x, y)$ at any point (x, y) in the region whose distance from the boundary is greater than ϵ (by the mean value property). Thus

$$|\nabla u| \leq \frac{C}{\epsilon}$$

or

$$|\nabla u| \cdot (\text{distance to boundary}) \leq C.$$

In our case, the distance to the boundary is comparable to y. So

$$y|\nabla u| \leq C_{\alpha,\beta,h,k}.$$

That proves the lemma.

$\square$

Now the almost everywhere finiteness of $S(u)$ on E will surely follow from

$$\int_E S^2(u)(x)\,dx < \infty$$

so we seek to prove this. Set

$$\mathcal{R} = \bigcup_{x \in E} \Gamma_\alpha^h(x).$$

So

$$\int_E S^2(u)(x)\,dx \le C \int_{\mathcal{R}} y|\nabla u|^2\,dxdy.$$

This follows by the already familiar device used in the global theorem.

With questions of convergence and smoothness set aside, a version of Green's theorem say that

$$\int_{\mathcal{R}} A\Delta B - B\Delta A\,dxdy = \int_{\partial\mathcal{R}} \left(A\frac{\partial B}{\partial \nu} - B\frac{\partial A}{\partial \nu} s \right)\,d\sigma,$$

where $\partial/\partial \nu$ is the normal derivative at $\partial\mathcal{R}$. We use this identity with

$$A = y,\quad \Delta A = 0,\quad B = \frac{|u|^2}{2},\quad \Delta B = |\nabla u|^2.$$

So we have

$$\int_{\mathcal{R}} y|\nabla u|^2\,dxdy = \frac{1}{2}\int_{\partial\mathcal{R}} \left(y\frac{\partial u^2}{\partial \nu} - u^2\frac{\partial y}{\partial \nu} \right)\,d\sigma, \qquad (*)$$

where $d\sigma$ is the element of area on $\partial\mathcal{R}$. Note that, since $\partial\mathcal{R}$ is composed of cones of fixed slope over $^c E$, then by an argument like the one used on the disc, $\partial\mathcal{R}$ has finite area.

Now

$$|(*)| \le \frac{1}{2}\int_{\partial\mathcal{R}} y\left|\frac{\partial u^2}{\partial \nu}\right|\,d\sigma + \frac{1}{2}\int_{\partial\mathcal{R}} |u|^2\left|\frac{\partial y}{\partial \nu}\right|\,d\sigma \equiv T_1 + T_2.$$

We have that

$$T_2 \le \int_{\partial\mathcal{R}} d\sigma < \infty.$$

Also,

$$T_1 \leq \int_{\partial \mathcal{R}} y|u||\nabla u|\, d\sigma \leq \int_{\partial \mathcal{R}} y|\nabla u|\, d\sigma \leq C \int_{\partial \mathcal{R}} d\sigma < \infty.$$

The last half-page or so has been a bit footloose. In order to make it precise, we need to overcome the fact that $\partial \mathcal{R}$ is jagged and the functions are not well defined when $y = 0$ (which is part of $\partial \mathcal{R}$ after all). So we define approximating regions on which the estimates hold uniformly.

Let

$$\partial \mathcal{R} = \mathcal{B} = \mathcal{B}^0 \cup \mathcal{B}^1,$$

where

$$\mathcal{B}^0 = \left\{ (x, y) : \frac{d(x, E)}{\alpha} = y, 0 < y < h \right\} = \{(x, y) \in \partial \mathcal{R} : 0 < y < h\}$$

and

$$\mathcal{B}^1 = \{(x, y) \in \partial \mathcal{R} : y = h\}.$$

We claim that there exist regions $\{\mathcal{R}_\epsilon\}$ such that

(i) $\overline{\mathcal{R}_\epsilon} \subseteq \mathcal{R}.$
(ii) $R_{\epsilon_1} \subseteq R_{\epsilon_2}$ if $\epsilon_2 < \epsilon_1.$
(iii) $\cup_\epsilon \mathcal{R}_\epsilon = \mathcal{R}.$
(iv) $\partial \mathcal{R}_\epsilon = \mathcal{B}_\epsilon^0 + \mathcal{B}_\epsilon^1$ with $\mathcal{B}_\epsilon^j$ similar to the above.

Let $\delta(x) = d(x, E)$ for $x \in \mathbb{R}^n$. Of course

$$|\delta(x) - \delta(x')| \leq |x - x'|$$

for all $x, x' \in \mathbb{R}^n$.

Let φ be a C_c^∞ function with $\int \varphi\, dx = 1$, $\varphi \geq 0$. Define $\varphi_\epsilon(x) = \epsilon^{-n}\varphi(x/\epsilon)$. Set $\delta_\epsilon(x) = \delta * \varphi_\epsilon(x)$. Then $|\delta_\epsilon(x) - \delta_\epsilon(x')| \leq |x - x'|$ uniformly in ϵ. And of course each δ_ϵ is C^∞. Also $\delta_\epsilon \to \delta$ uniformly. Choose $c_\epsilon > 0$ so that the functions $\delta_\epsilon + c_\epsilon$ decrease monotonically to δ.

Define

$$B_\epsilon^0 = \left\{ (x, y) : y = \frac{\delta_\epsilon + c_\epsilon}{\alpha} \right\}$$

and

$$B_\epsilon^1 = \{(x, y) : y = h - \epsilon\}.$$

Refer to Fig. 5.4.

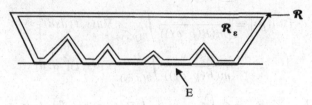

Fig. 5.4 The sets B_ϵ^0 and B_ϵ^1

We would like to see that

$$\iint_{\mathcal{R}_\epsilon} y|\nabla u|^2\, dx\, dy \leq C < \infty$$

with C independent of ϵ. Now Green's theorem says, as before, that

$$\int_{\mathcal{R}_\epsilon} y|\nabla u|^2\, dx\, dy = \frac{1}{2}\int_{\partial \mathcal{R}_\epsilon} y\frac{\partial u^2}{\partial \eta_\epsilon}\, d\sigma_\epsilon - \frac{1}{2}\int_{\partial \mathcal{R}_\epsilon} u^2 \frac{\partial y}{\partial \eta_\epsilon}\, d\sigma_\epsilon\,,$$

where η_ϵ, $\mathcal{R}_\epsilon$, and $d\sigma_\epsilon$ are related in the obvious ways.

Now, for $\epsilon < h/3$, the integrals over B_ϵ^1 are estimated as before. On the other hand, B_ϵ^0 is locally the graph of a function $y = \varphi_\epsilon(x)$ and

$$d\sigma_\epsilon = \sqrt{1 + |\nabla_x \varphi_\epsilon(x)|^2}\,.$$

But $|\nabla_x \varphi| \leq C_\alpha$ so $d\sigma_\epsilon \leq C\, dx$. It follows that $d\sigma_\epsilon \approx dx$ uniformly in ϵ. So the integrals over B_ϵ^0 are estimated as before provided we recall that $|u| \leq 1$ and $y|\nabla u| \leq C$. The apparent dependence of this last estimate on the boundedness of u on Γ_β^k strictly larger than Γ_α^h is removed by the usual density argument.

CONVERSE In order to prove the converse, we endeavor to unravel the above. We first try to show that $u\big|_{\partial \mathcal{R}_\epsilon}$ is in L^2 of this boundary. So we suppose that the situation has been uniformized so that, for all $x_0 \in E_0$ a measurable subset of $\mathbb{R}^n$ and for some fixed β, k positive we have

$$\int_{\Gamma_\beta^k(x_0)} |\nabla u|^2 y^{1-n}\, dx\, dy \leq 1\,.$$

Lemma 5.5.5 *Under the above conditions, $|y\nabla u| \leq C$ in a smaller cone.*

Proof Fix $\alpha < \beta$ and $h < k$. Then, for all $p = (x, y) \in \Gamma_\alpha^h(x_0)$, there exists $c > 0$ such that $B(p, cy) \subseteq \Gamma_\beta^k$. Now use the mean value property:

$$|\nabla u(p)| = \left| \frac{1}{m(B(p,cy))} \int_{B(p,cy)} \nabla u(s,t) \, dsdt \right|$$

$$\leq \frac{1}{m(B(p,cy))} \int_{B(p,cy)} |\nabla u(s,t)|^2 \, dsdt$$

$$\leq C \int_{B(p,cy)} y^{-1} \cdot \left(|\nabla u(s,t)|^2 y^{1-n} \right) dsdt$$

$$\leq y^{-1} \cdot C \cdot \int_{\Gamma_\beta^k(x_0)} |\nabla u(s,t)|^2 t^{1-n} \, dsdt$$

$$\leq C \cdot y^{-1}$$

or $y|\nabla u(x,y)| \leq C$ as desired.   □

Now choose $E_0 \subseteq E$, E_0 compact, with $m(E \setminus E_0)$ small and so that

$$\frac{m(E_0 \cap B(x,r))}{m(B(x,r))} \geq \frac{1}{2}$$

for all $x \in E_0$, $r < \eta$ constant. Let

$$\mathcal{R} = \bigcup_{x_0 \in E} \Gamma_\alpha^h(x_0).$$

We claim that

$$\int_{\mathcal{R}} y|\nabla u|^2 \, dxdy < \infty.$$

Let

$$\psi(x_0, x, y) = \begin{cases} 1 & \text{if } |x - x_0| < \beta y, \ 0 < y < k \\ 0 & \text{otherwise}. \end{cases}$$

Then

$$\infty > \int_{E_0} \int_{\Gamma_\beta^k(x_0)} y^{1-n} |\nabla u|^2 \, dxdydx_0$$

$$= \iiint_{E_0} \psi(x_0, x, y) \, dx_0 y^{1-n} |\nabla u|^2 \, dxdy.$$

It will suffice to see that

$$\int_{E_0} \psi(x_0, x, y) dx_0 \geq C y^n$$

whenever $(x, y) \in \mathcal{R}$.

If $(x, y) \in \mathcal{R}$, then there exists a $z \in E$ such that $|x - z| < \alpha y$, $0 < y < h$. Thus

$$\int_{E_0} \psi(x_0, x, y) dx_0 \geq \int_{E_0 \cap \{|x_0 - z| < (\beta - \alpha)y\}} dx_0$$

$$\geq m(E \cap \{|x_0 - z| < (\beta - \alpha)y\})$$

$$\geq c y^n$$

by the density property we assumed for E. Thus

$$\infty > \int_E S(u)(x_0) \, dx_0$$

$$\geq \int_{\mathcal{R}} y|\nabla u|^2 \, dx dy$$

$$\geq \int_{\mathcal{R}_\epsilon} y|\nabla u|^2 \, dx dy$$

$$\overset{\text{(Green)}}{\approx} C \left| \int_{\mathcal{B}_\epsilon} y \frac{\partial u^2}{\partial \eta_\epsilon} - u^2 \frac{\partial y}{\partial \eta_\epsilon} d\sigma_\epsilon \right|. \tag{*}$$

Here $\mathcal{B}_\epsilon \equiv \mathcal{B}_\epsilon^0 \cup \mathcal{B}_\epsilon^1$. Note that the points on $\mathcal{B}_\epsilon^1$ are bounded from $\{y = 0\}$ hence the contribution to $(*)$ from $\mathcal{B}_\epsilon^0$ is bounded.

Thus we have

$$\infty > \int_{\mathcal{B}_\epsilon^0} y \frac{\partial u^2}{\partial \eta_\epsilon} - u^2 \frac{\partial y}{\partial \eta_\epsilon} d\sigma_\epsilon.$$

Now we claim that

$$-\frac{\partial y}{\partial \eta_\epsilon} \geq C > 0$$

on $\mathcal{B}_\epsilon^0$. In fact $\partial/\partial \eta_\epsilon$ is the outward unit normal to the surface defined by $\varphi_\epsilon(x, y) = \alpha y - \delta_\epsilon(x) \equiv 0$. Thus $\partial/\partial \eta_\epsilon$ is given by

$$\left(-\frac{\partial \varphi_\epsilon}{\partial y}, \frac{\partial \varphi_\epsilon}{\partial x_1}, \ldots, \frac{\partial \varphi_\epsilon}{\partial x_n} \right) = \left(-\alpha, \frac{\partial \delta_\epsilon}{\partial x_1}, \ldots, \frac{\partial \delta_\epsilon}{\partial x_n} \right)$$

normalize to have length 1. So

$$\frac{\partial y}{\partial \eta_\epsilon} \leq -\alpha(\alpha^2 + n^2)^{-1/2}$$

or

$$-\frac{\partial y}{\partial \eta_\epsilon} \geq \alpha(\alpha^2 + n^2)^{-1/2},$$

as desired.

So we know that

$$C_1 \int_{B_\epsilon^0} \left(\frac{y \partial u^2}{\partial \eta_\epsilon} - u^2 \frac{\partial y}{\partial \eta_\epsilon} \right) d\sigma_\epsilon \leq C_2$$

or

$$C_1 \int_{B_\epsilon^0} -\frac{\partial y}{\partial \eta_\epsilon} u^2 d\sigma_\epsilon \leq C_2 + C_1 \left| \int_{B_\epsilon^0} \frac{y \partial u^2}{\partial \eta_\epsilon} \right|$$

$$\leq C_2 + C_1 \int_{B_\epsilon^0} y |u| |\nabla u| \, d\sigma_\epsilon$$

$$\leq C_2 + C_3 \int_{B_\epsilon^0} |u| \, d\sigma_\epsilon. \tag{$\star$}$$

Let

$$\mathcal{J}_\epsilon = \left(\int_{B_\epsilon^0} u^2 \, d\sigma_\epsilon \right)^{1/2}.$$

Then $(\star)$ just says that

$$C_4 \mathcal{J}_\epsilon^2 = C_4 \int_{B_\epsilon^0} u^2 \, d\sigma_\epsilon$$

$$\leq C_2 + C_3 \int_{B_\epsilon^0} |u| \, d\sigma_\epsilon$$

$$\leq C_2 + C_3 \int_{B_\epsilon^0} |u|^2 \, d\sigma_\epsilon^{1/2} \cdot \left(\int_{B_\epsilon^0} d\sigma_\epsilon \right)^{1/2}$$

$$\leq C_2 + C_5 \mathcal{J}_\epsilon,$$

since B_ϵ^0 is rectifiable. It follows that $\mathcal{J}_\epsilon \leq C_6$, with C_6 independent of ϵ. Thus $\int_{B_\epsilon^0} u^2 \, d\sigma_\epsilon \leq C$. This is what we set to prove a while back. Now we complete the proof of the converse.

We wish to see that u is nontangentially bounded almost everywhere on E. We will find a V harmonic satisfying the following:

(i) $V = PI(f)$ with $f \in L^2(\mathbb{R}^n)$, $f \geq 0$.
(ii) $|u(x, y)| \leq C \cdot V(x, y) + C_1$ for $(x, y) \in \mathcal{R}$.

By the local Fatou theorem, this will do the trick. We use what is at hand to construct V.

For each $\epsilon > 0$, set

$$f_\epsilon(x) = \begin{cases} |u(x, \delta_\epsilon(x)/\alpha)| & \text{if } (x, \delta_\epsilon(x)/\alpha) \in B_\epsilon^0 \\ 0 & \text{otherwise}. \end{cases}$$

Recall here that $B_\epsilon^0 = \{y = \delta_\epsilon(x)/\alpha : 0 < y < h - \epsilon\}$.

Trivially, $d\sigma_\epsilon \geq dx$ so that

$$\int_{\mathbb{R}^n} |f_\epsilon(x)|^2\, dx \leq \int_{B_\epsilon^0} |u(x)|^2\, d\sigma_\epsilon \leq C$$

by the estimate we have proved above.

Let $v_\epsilon(x, y) = PI(f_\epsilon)(x)$. Our main claim is that, for appropriate $C > 0$,

$$|u(x, y)| \leq C(v_\epsilon(x, y) + 1) \quad \text{for } (x, y) \in \mathcal{R}_\epsilon. \qquad (\ast\ast\ast)$$

By the maximum principle, it suffices to check this inequality on $\partial\mathcal{R}_\epsilon$. The non-critical part of $\partial\mathcal{R}_\epsilon$ (bounded away from $\{y = 0\}$ and compact) is taken care of by making C large enough, in the $C \cdot 1$ term. So we check $(\ast\ast\ast)$ on B_ϵ^0.

Now $B_\epsilon^0 \subseteq \mathcal{R} = \cup_{x_0 \in E} \Gamma_\alpha^h(x_0)$ so that there exists a constant c with $1/2 > c > 0$ such that $B((x, y), cy) \subseteq \cup_{x_0 \in E} \Gamma_{\alpha^*}^{h^*}$ for all $(x, y) \in B_\epsilon^0$. Here $\alpha < \alpha^* < \beta$ and $h < h^* < k$ fixed. Fix $p_1 = (x, y) \in B_\epsilon^0$ and let p_2 be any other point of $B = B((x, y), cy)$. Then

$$|u(p_1) - u(p_2)| \leq |p_1 - p_2| \cdot \sup_{(s,t) \in \overline{p_1 p_2}} |\nabla u(s, t)|$$

$$\leq c \cdot y \cdot \sup_{y/2 < |t| < (3/2)y} |\nabla u(s, t)|$$

$$\leq c$$

since $|y \nabla u| \leq c$.

We conclude that $|u(p_1)| \leq |u(p_2)| + c$. Let

$$|S_\epsilon| = \int_{B_\epsilon^0 \cap B} d\sigma_\epsilon.$$

Then, taking the mean value in the p_2 variable gives

$$|u(p_1)| \leq \frac{1}{|S_\epsilon|} \int_{S_\epsilon} |u_\epsilon(p_2)| \, d\sigma_\epsilon(p_2) + C.$$

Clearly there exists a constant A such that $|S_\epsilon| \geq Ay^n$. Hence we have

$$|u(p_1)| \leq Cy^{-n} \int_{|z-x|<cy} |f_\epsilon(z)| \, dz + C_2$$

if we recall the definition of f_ϵ and the fact that $dx \geq c\,d\sigma_\epsilon$.

Finally, since it is trivial that $P_y(z) \geq ky^{-n}$ for $|z| < cy$, we have

$$|u(p_1)| \leq cv_\epsilon(x, y) + C_2 \text{ for } p_1 \in \mathcal{R}_\epsilon.$$

Since the f_ϵ are uniformly L^2 bounded, select a weak-$*$ convergent subsequence $f_{\epsilon_k} \xrightarrow{w^*} f$. Thus $v_{\epsilon_k} \to v$ pointwise, where $v = PI(f)$.

Since $\mathcal{R}_\epsilon \to \mathcal{R}$ we obtain

$$|u(x, y)| \leq Cv(x, y) + C_2 \text{ for } (x, y) \in \mathcal{R}.$$

It follows that u is nontangentially bounded almost everywhere on E hence has nontangential limits almost everywhere on E. Letting $E \to E_0$ and justifying the uniformization by a density argument completes the proof. $\qquad\square$

We would like to obtain a new global analogue of the above, valid for $0 < p < \infty$. We need some definitions. Fix $\alpha > 0$ and let $u(x, y)$ be given on $\mathbb{R}_+^{n+1}$. Set

$$u_\alpha^*(x) = \sup_{(x',y')\in\Gamma_\alpha(x)} |u(x', y')| = u^*(x).$$

Set

$$S_\beta^2 u(x) = \iint_{\Gamma_\beta(x)} |\nabla u(x', y')|^2 y^{1-n} \, dx'dy' = S^2 u(x).$$

Theorem 5.5.6

(a) *Suppose that u is harmonic on $\mathbb{R}_+^{n+1}$ and $u^* \in L^p$. Then $\|Su\|_{L^p} \leq C_p\|u^*\|_{L^p}$, $0 < p < \infty$.*

(b) *If $u \in h(\mathbb{R}_+^{n+1})$, suppose that $u(x, y) \to 0$ as $y \to \infty$. Then, if $S(u) \in L^p$, then $u^* \in L^p$ and $\|u^*\|_{L^p} \leq C_p\|S(u)\|_{L^p}$, $0 < p < \infty$.*

This theorem has a long history. There is a four-fold decomposition of the problem:

(1) $p \geq 1$.

(2) $0 < p < 1$.

(3) $n = 1$.

(4) general n.

The first proof, in case $n = 1$, was by Burkholder–Gundy–Silverstein using Brownian motion. In order to prove the theorem, we will need to make some of the last proof quantitative.

We will use the following elementary fact about distribution functions: Let f be given on $\mathbb{R}^n$. For $\alpha > 0$, define

$$\lambda_f(\alpha) = m\{x : |f(x)| > \alpha\} \ , \ \alpha > 0.$$

Then

$$\int_{\mathbb{R}^n} |f(x)|^p \, dx = \int_0^\infty \alpha^p d\lambda(\alpha) \stackrel{\text{(parts)}}{=} p \cdot \int_0^\infty \lambda(\alpha)\alpha^{p-1} \, d\alpha \, .$$

Lemma 5.5.7 *Suppose that $u(x, y)$ is continuous on $\mathbb{R}^{n+1}_+$. Then, for $\gamma > 0$, we have*

$$m\{x : u^*_{\beta_1}(x) > \gamma\} \le C_{\beta_1,\beta_2} m\{x : u_{\beta_2} * (x) > \gamma\}$$

independent of $\gamma > 0$, any $\beta_1, \beta_2 > 0$.

The essence of this lemma is that $\|u^*_{\beta_1}\|_{L^p} \approx \|u^*_{\beta_2}\|_{L^p}$.

Proof of the Lemma This is only interesting when $\beta_1 > \beta_2$. Now fix γ temporarily. Let $E = \{u^*_{\beta_2} > \gamma\}$. Let χ_E be the characteristic function of E. Set $E^* = \{x : M\chi_E(x) > C\}$ for a fixed constant C and M the Hardy–Littlewood maximal function. Of course

$$m(E^*) \le \frac{5^n \cdot m(E)}{C} \, .$$

We claim that

$$\{x : u^*_{\beta_1} > \gamma\} \subseteq E^*$$

for the right choice of C. This is because $u^*_{\beta_1}(x) > \gamma$ means that there exists a $p = (a, b) \in \Gamma_{\beta_1}(x)$ such that $|u(p)| > \gamma$. See Fig. 5.5.

But then $u^*_{\beta_2}(t) > \gamma$ for all t in the upside down cone at (a, b), that is all t so that $|t - a| < \beta_2 b$. But then

$$\frac{m(E \cap B(x, |a - x| + \beta_2 b)}{m(B(x, |a - x| + \beta_2 b)} \ge \frac{m(B(a, \beta_2 b)}{m(B(x, |a - x| + \beta_2 b))}$$

$$\ge \frac{(\beta_2 b)^n}{(\beta_1 b + \beta_2 b)^n}$$

$$\ge \frac{\beta_2^n}{(\beta_1 + \beta_2)^n} \, .$$

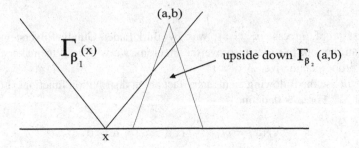

Fig. 5.5 The approach region $\Gamma_{\beta_1}(x)$

Thus if we let

$$C = \frac{\beta_2^n}{(\beta_1 + \beta_2)^n},$$

then we find that $p \in \{x : M\chi_E(x) > C\}$ so that

$$m\{x : u_{\beta_1}^* > \gamma\} \le \frac{5^n \cdot (\beta_1 + \beta_2)^n}{\beta_2^n} \cdot m\{x : u_{\beta_2}^*(x) > \gamma\}.$$

That proves the result.   □

A reference for this work is Fefferman and Stein [FES].

Now we prove the theorem. We restrict attention to $p < 2$. The case $p \ge 2$ is quite different, has been known longer, and can be handled with different techniques. It can also be handled by a more elaborate version of the same method, c.f. [STE1].

We will suppose that the situation has been uniformized in the following way:

(1) $u = PI(f)$, $f \in L^2$.
(2) The cone which defines u^* is strictly larger than that which defines $S(u)$. Say they are of apertures $\beta > \alpha$.

Taking inspiration from the lemma, we estimate that $m\{x : S(u) > \lambda\}$ in terms of $m\{x : u^*(x) > \lambda\}$. Let

$$E = \{x : u^*(x) \le \lambda\} \qquad \text{and} \qquad B = {}^cE = \{x : u^*(x) > \lambda\}.$$

Then E is closed and B is open. Let $\mathcal{R} = \cup_{x \in E}\Gamma_\alpha(x)$.

As usual, we have

$$\int_E S^2(u)(x)\,dx \le \iint_{\mathcal{R}} y|\nabla u|^2\,dxdy = \lim_{\epsilon \to 0} \iint_{\mathcal{R}_\epsilon} y|\nabla u|^2\,dxdy.$$

Fig. 5.6 The boundary sets $\mathcal{B}_\epsilon^E$ and $\mathcal{B}_\epsilon^B$

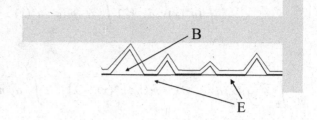

Again we introduce the approximating regions so that it is legitimate to apply Green's theorem. But

$$\iint_{\mathcal{R}_\epsilon} y|\nabla u|^2 \, dx dy \overset{\text{(Green)}}{=} \frac{1}{2} \int_{\partial \mathcal{R}_\epsilon} y \frac{\partial u^2}{\partial \eta_\epsilon} \, d\sigma_\epsilon - \frac{1}{2} \int_{\partial \mathcal{R}_\epsilon} \frac{\partial y}{\partial \eta_\epsilon} u^2 \, d\sigma_\epsilon . \qquad (*)$$

Now since we have assumed that $u = PI(f)$, $f \in L^2$, we have $\sup_y |u(x, y)| \leq AMf(x) \in L^2$. In addition, since the kernel of $y\nabla u$ is $y \cdot \nabla P_y$, which is essentially an approximation to the identity, we have

$$\sup_y y|\nabla u| \in L^2 . \qquad (\star)$$

Also $y|\nabla u(x, y)| \to 0$ almost everywhere since this is easy to check for $f \in C_c$ and $(\star)$ tells us that the associated maximal operator is bounded.

Write

$$\partial \mathcal{R}_\epsilon = \mathcal{B}_\epsilon = \mathcal{B}_\epsilon^E \cup \mathcal{B}_\epsilon^B ,$$

where of course $\mathcal{B}_\epsilon^E$ is the portion which lies over E and $\mathcal{B}_\epsilon^B$ is the portion which lies over B. See Fig. 5.6.

But over $\mathcal{B}_\epsilon^B$ we know that $|u| \leq \lambda$ so $|y\nabla u| \leq C\lambda$. Thus the contribution to $(*)$ from $\mathcal{B}_\epsilon^B$ is $\leq C\lambda^2 m(B)$. Therefore

$$\int_E S^2(u)(x) \, dx \leq C\lambda^2 m\{x : u^*(x) > \lambda\} + \left| \int_{\mathcal{B}_\epsilon^E} y \frac{\partial u^2}{\partial \eta_\epsilon} \, d\sigma_\epsilon \right| + \left| \int_{\mathcal{B}_\epsilon^E} \frac{\partial y}{\partial \eta_\epsilon} u^2 \, d\sigma_\epsilon \right| .$$

Since $y\nabla u \to 0$ as $y \to 0$, the second term will die with ϵ. The last term is dominated by $\int_E |u^*|^2 \, dx$.

Altogether then,

$$\int_E S^2(u)(x) \, dx \leq C\lambda^2 m\{x : u^*(x) > \lambda\} + \int_E |u^*|^2 \, dx .$$

Clearly

$$\int_E (u^*(x))^2 \, dx \leq 2 \int_0^\lambda \alpha \cdot m\{x : u^*(x) > \alpha\} \, d\alpha \, .$$

So

$$\int_E S^2(u)(x) \, dx \leq C\lambda^2 \cdot m\{x : u^*(x) > \lambda\} + 2 \int_0^\lambda \alpha \cdot m\{x : u^*(x) > \alpha\} \, d\alpha \, .$$

Now

$$m\left(\{x : |S(u)(x)| > \lambda\} \cap E\right) \leq \frac{1}{\lambda^2} \int_E S^2(u)(x) \, dx \, ,$$

so

$$m\{x : |S(u)(x)| > \lambda\} \leq \frac{1}{\lambda^2} \int_E S^2(u)(x) \, dx + m(^cE) \, . \qquad (**)$$

But

$$m(^cE) = m(B) = m\{u^* > \lambda\} \, .$$

Altogether then,

$$m\{S(u) > \lambda\} \leq C \cdot m\{u^* > \lambda\} + \frac{C}{\lambda^2} \int_0^\lambda \alpha \cdot m\{u^* > \alpha\} \, d\alpha \, .$$

Thus

$$\int S(u)^p \, dx = p \int_0^\infty \lambda^{p-1} \cdot m\{Su > \lambda\} \, d\lambda$$

$$\leq C_p \int_0^\infty \lambda^{p-1} m\{u^* > \lambda\} \, d\lambda + C_p \int_0^\infty \lambda^{p-3} \int_0^\lambda \alpha \cdot m\{u^* > \alpha\} \, d\alpha d\lambda$$

$$\leq C\|u^*\|_{L^p} + C_p \int_0^\infty \int_\alpha^\infty \lambda^{p-3} \, d\lambda \alpha \cdot m\{u^* > \alpha\} \, d\alpha \, .$$

For $p < 2$, the second integral in λ converges at ∞ so this is

$$\leq C\|u^*\|_{L^p} + \frac{C_p}{2-p} \int_0^\infty \alpha^{p-1} m\{u^* > \alpha\} \, d\alpha$$

$$\leq C_p\|u^*\|_{L^p} \, .$$

This completes the proof.   □

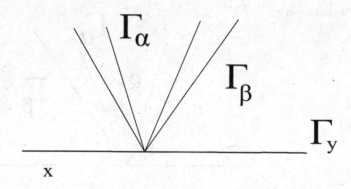

Fig. 5.7 Comparison of cones

We omit the proof of the converse, the removal of the condition $u = PIf$, $f \in L^2$. The lemma clears up the cone condition. Note that the assumption $u = PIf$ was used to control u at ∞ so that we could apply Green's theorem. Burkholder and Gundy have a proof of the above result in 1972 in *Studia*. We will only use this result for $p \leq 1$.

5.6 Applications of the Maximal Function Characterization

Suppose that u and v are conjugate harmonic functions on $\mathbb{R}^2_+$. Hence $|\nabla u|^2 = |\nabla v|^2$. The above theorem enables us to pass from u to v or, more precisely, from u^* to v^*. This is because

$$\|u^*\|_{L^p} \leq C_p \|Su\|_{L^p} = C_p \|Sv\|_{L^p} \leq C_p \|v^*\|_{L^p}$$

and conversely. In higher dimensions, we have $u, v_1, \ldots, v_n$ with

$$\frac{\partial u}{\partial x_j} = \frac{\partial v_j}{\partial y}.$$

There is no simple relationship between $\operatorname{grad} u$ and $\operatorname{grad} v_j$. So we need an additional tool to relate u, v_j via the area integral. See Fig. 5.7.

Lemma 5.6.1 *Let u be harmonic in $\mathbb{R}^{n+1}_+$ and bounded on some halfspace $y \geq y_0$. Then*

$$\int_{\Gamma_\alpha} y^{1-n} \left| \frac{\partial u}{\partial y} \right|^2 dxdy \leq C_{\alpha,\beta} \int_{\Gamma_\beta} y^{1-n} \left| \frac{\partial u}{\partial x_j} \right|^2 dxdy.$$

Fig. 5.8 The vector ρ in the cone Γ_β

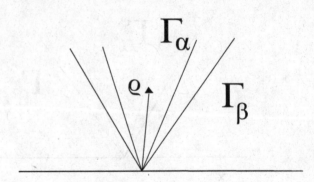

[We assume here that $\partial u/\partial x_j \to 0$ as $y \to \infty$ in the cone Γ_β.]

The motivation for the lemma is that, if we know the area integral for u, then we know $\partial u/\partial x_j$, hence we know $\partial v_j/\partial y$ and, by the lemma, we therefore know $\partial v_j/\partial x_k$. We note that, for $n = 1$, it suffices to let $\alpha = \beta$. It is unknown whether this can be done for $n > 1$.

Outline of the Proof of the Lemma Let ρ be a unit vector in Γ_α. We prove the lemma for each ρ, then integrate out over ρ. Refer to Fig. 5.8 as we proceed.

So we will prove that

$$\int_0^\infty \left| \left(\frac{\partial u}{\partial x_j} \right)(\rho s) \right|^2 s \, ds \le C_{\alpha,\beta} \int_{\Gamma_\beta} \left| \frac{\partial u}{\partial y} \right|^2 y^{1-n} \, dx dy.$$

As long as $\rho \in \Gamma_\alpha$, then s and the y component of ρs are comparable. So, integrating both sides of the inequality over $S^n \cap \Gamma_\alpha$, yields the result. Now fix ρ once and for all. Write

$$\frac{\partial u(x, y)}{\partial x_j} = -\int_y^\infty \frac{\partial^2 u(x, \tau)}{\partial y \partial x_k} \, d\tau.$$

The integral converges since u is bounded. As usual, there exists $C > 0$ such that $(x, \tau) \in \Gamma_\alpha$ implies that $B((x, \tau), C\tau) \subseteq \Gamma_\beta$. Call this ball B_τ.

Clearly B_τ is contained in the strip $S_\tau = \{(1 - C)\tau \le y \le (1 + C)\tau\}$. We claim that

$$\left| \frac{\partial^2 u(x, \tau)}{\partial y \partial x_k} \right|^2 \le C\tau^{-2} \frac{1}{m(B_\tau)} \int_{B_\tau} \left| \frac{\partial u(x', y')}{\partial y} \right|^2 dx' dy'$$

$$= C\tau^{-n-3} \int_{B_\tau} \left| \frac{\partial u(x', y')}{\partial y} \right|^2 dx' dy'. \tag{*}$$

This follows by convolving u with a radial bump function φ_τ such that $\varphi \ge 0$, $\text{supp}\,\varphi \subseteq B(0, 1/2)$, $\int \varphi = 1$, $\varphi \in C^\infty$. Harmonicity implies that $u * \varphi = u$ and differentiation under the integral sign yields (∗). Thus

$$\left|\frac{\partial^2 u(x,\tau)}{\partial y \partial x_k}\right|^2 \leq C \cdot \tau^{-n-3} \int_{S_\tau \cap \Gamma_\beta} \left|\frac{\partial u}{\partial y}\right|^2 dx' dy' .$$

In conclusion,

$$\left|\frac{\partial u}{\partial x_j}(\rho s)\right| \leq C \int_{cs}^\infty \tau^{(-n-3)/2} I_\tau^{1/2} d\tau ; \qquad (\star)$$

where

$$I_\tau = \int_{S_\tau \cap \Gamma_\beta} \left|\frac{\partial u}{\partial y}\right|^2 dx' dy' .$$

We need a special case of Hardy's inequality:

(a) If $\Phi(s) = \int_s^\infty \varphi(\tau) d\tau$, then

$$\int_0^\infty |\Phi(s)|^2 s \, ds \leq C \int_0^\infty |\varphi(\tau)|^2 \tau^3 d\tau .$$

Proof We see that

$$|\Phi(s)| \leq \int_s^\infty |\varphi(\tau)| \cdot \tau \cdot \tau^{-1} d\tau$$

so, by Schwarz,

$$|\Phi(s)|^2 \leq \int_s^\infty |\varphi(\tau)|^2 \tau^2 d\tau \cdot \int_s^\infty \frac{1}{\tau^2} d\tau$$

$$\leq \int_s^\infty |\varphi(\tau)|^2 \tau^2 d\tau \cdot \frac{1}{s}$$

or

$$\int_0^\infty |\Phi(s)|^2 s \, ds \leq \int_0^\infty dx \int_s^\infty |\varphi(\tau)|^2 \tau^2 d\tau$$

$$= \iint\limits_{0 \leq s \leq \tau} |\varphi(\tau)|^2 \tau^2 d\tau$$

$$= \frac{1}{2} \int_0^\infty |\varphi(\tau)|^2 \tau^3 d\tau .$$

That proves Hardy's inequality.   □

Applying this last result to ($\star$) we get

$$\int_0^\infty \left|\frac{\partial u}{\partial x_j}(\rho s)\right|^2 s\, ds \leq C \int_0^\infty \tau^{-n-3} \cdot I_\tau \cdot \tau^3\, d\tau$$

$$= C \int_0^\infty \int_{S_\tau \cap \Gamma_\beta} \left|\frac{\partial u}{\partial y}(x', y')\right|^2 dx'dy'\tau^{-n}\, d\tau$$

$$\leq C \int_{\Gamma_\beta} \left|\frac{\partial u}{\partial y}(x', y')\right|^2 \cdot |y'|^{1-n}\, dx'dy',$$

as desired.

We will have considerable use for the following fact:

Proposition 5.6.2 *If* $u \in H^p$ *(in the sense that* $(u, R^1 u, \ldots, R^n u) \in H^p$*),* $0 < p < \infty$*, then* $u^* \in L^p$ *and* $\|u^*\|_{L^p} \leq A_p \|u\|_{H^p}$.

Proof We do the proof for $p > (n-1)/n$ only. Note that the case $p > 1$ is obvious. Now we know that

$$\left|\sum |u_j(x, y + \epsilon)|^2\right|^{(n-1)/(2n)}$$

is subharmonic. Therefore

$$\left(\sum |u_j(x, y + \epsilon)|^2\right)^{(n-1)/(2n)} \leq \int P_y(x - t) \left(\sum |u_j(t, \epsilon)|^2\right)^{(n-1)/(2n)} dt$$

since the left side is subharmonic and the right side harmonic. But the $(\sum |u_j(t, \epsilon)|^2)^{(n-1)/(2n)}$ are uniformly bounded in $L^{p/((n-1)/n)}$ so there exists a weak-$*$ subsequence converging to some $h(x)$. If $h(x, y) = PIh$, then it follows that

$$\left(\sum |u_j(x, y)|^2\right)^{(n-1)/(2n)} \leq h(x, y)$$

and

$$\sup_{y>0} \int_{\mathbb{R}^n} h(x, y)^{p/((n-1)/n)}\, dx = \|h\|_{L^q(\mathbb{R}^n)}^q = \|u\|_{H^p}^p,$$

where we let $q = p/((n-1)/n) > 1$. So we have

$$u^*(x) \leq h^*(x)^{n/(n-1)} \leq C(Mh(x))^{n/(n-1)},$$

where M is the Hardy–Littlewood maximal operator. So

$$\int_{\mathbb{R}^n} |u^*(x)|^p \, dx \le C \int_{\mathbb{R}^n} |Mh(x)|^q \, dx \le C \int_{\mathbb{R}^n} h^q \, dx = \|u\|_{H^p}^p,$$

as desired. □

Before continuing, we present an outline of where we are going and what the remainder of these notes will contain.

I. Real Variable Theory: Let us consider $H^p(\mathbb{R}^n)$. Here we can characterize H^p without appealing to any notion of conjugate harmonic function.

II. Several Complex Variables:

  (a) The polydisc. One of the key features is the presence of a distinguished boundary, different from the topological boundary.
  (b) The unit ball in $\mathbb{C}^n$. The boundary and the distinguished boundary coincide. The boundary is smooth and pseudoconvex.
  (c) The domains of Cartan. These have some of the characteristics of **(a)** and **(b)**. The boundaries turn out to be "semi-distinguished."

Returning to the subject at hand, we proceed by giving a characterization of H^p. We begin with $H^p(D)$.

In what follows, T is a fixed triangle in D with vertex at 1 and T_θ is the triangle obtained by rotating T to have vertex at θ.

Theorem 5.6.3 *If $F \in H(D)$ with $F = u + iv$, u, v real, $v(0) = 0$, and if $u^*(\theta) = \sup_{z \in T_\theta} |u(z)|$, then $F \in H^p$ if and only if $u^* \in L^p$, $0 < p < \infty$, and in this case, $\|F\|_{H^p} \approx \|u^*\|_{L^p}$.*

Remark 5.6.4 This theorem of Burkholder, Gundy, and Silverstein was originally proved by means of Brownian motion and represents the first theorem of classical analysis which was proved by these methods.

In the classical setting, the area integral takes the form

$$S^2(u)(\theta) = \iint_{\Gamma_\theta} |\nabla u|^2(z) \, dx \, dy.$$

Then the *area theorem* for harmonic functions in D is this:

Theorem 5.6.5 *Let u be harmonic in D, $0 < p < \infty$. Then*

(1) *If $u^* \in L^p$, then $S(u) \in L^p$ and $\|Su\|_p \le C_p \|u^*\|_p$.*
(2) *If $u(0) = 0$ and $S(u) \in L^p$, then $u^* \in L^p$ and $\|u^*\|_p \le A_p \|Su\|_p$.*

Let us show how to obtain Theorem 5.6.3 from Theorem 5.6.5. Suppose that $u \in H(D)$ and $u^* \in L^p$. By the area theorem, $S(u) \in L^p$. Now the crucial point is that $S(u) = S(v)$ since $|\nabla u|^2 = |\nabla v|^2$. So $S(v) \in L^p$. By the area theorem, since $v(0) = 0$, $v^* \in L^p$ and $\|v^*\|_{L^p} \le A_p \|u^*\|_{L^p}$. (Here v is the conjugate of u.)

So $F = u + iv \in H^p$, $\|F\|_{H^p} \le A_p \|u^*\|_{L^p}$. So the central point is that the area integral is invariant under the Hilber transform.

We already know that $\|F^*\|_{L^p} \le A_p \|F\|_{H^p}$. This gives the converse.

Now we have a characterization of H^p independent of any notion of conjugate harmonic function. We shall seek another characterization of the following type: The class of distributions on the circle S which corresponds to the H^p functions on D are those distributions whose maximal function associated with any regularization are in L^p.

Now we can characterize $H^p(\mathbb{R}^{n+1}_+)$ in the same way we did $H^p(D)$. We have

Theorem 5.6.6 (1) *Let* $F = (u_0, u_1, \ldots, u_n)$, $u = u_0$, *be a system of conjugate harmonic functions in* $H^p(\mathbb{R}^{n+1}_+)$, *and* $p > (n-1)/n$. *Then* $u^* \in L^p$ *and* $\|u^*\|_p \le A_p \|F\|_p$.

(2) *Conversely, suppose that* u *is harmonic on* $\mathbb{R}^{n+1}_+$ *and* $u^* \in L^p$, $p > (n-1)/n$. *Then there exist harmonic functions* $u_1, \ldots, u_n$ *in* $\mathbb{R}^{n+1}_+$ *so that* $F = (u, u_1, \ldots, u_n) \in H^p(\mathbb{R}^{n+1}_+)$ *and* $\|F\|_{H^p} \le C_p \|u^*\|_{L^p}$.

Recall the area theorem for $\mathbb{R}^{n+1}_+$: If $u \in h(\mathbb{R}^{n+1}_+)$, then

(1) For $u^* \in L^p$, $\|Su\|_{L^p} \le A_p \|u^*\|_{L^p}$, $0 < p < \infty$.
(2) If $u(x, y) \to 0$ as $y \to \infty$ and $S(u) \in L^p$, then $u^* \in L^p$ and $\|u^*\|_{L^p} \le C_p \|S(u)\|_{L^p}$.

In order to prove this last theorem, we need the earlier lemma and also this coming lemma:

Lemma 5.6.7 (i) *Suppose that* $0 < p < \infty$ *and* $u \in h(\mathbb{R}^{n+1}_+)$ *satisfies* $\int_{\mathbb{R}^n} |u(x, y)|^p \, dx \le A$ *for all* $y > 0$. *Then* $|u(x, y)| = O(y^{-n/p})$.
(ii) *If* $u \in h(\mathbb{R}^{n+1}_+)$ *and* $|u(x, y)| \le Ay^{-\beta}$, *some* $\beta \in \mathbb{R}$, *then*

$$\left| \left(\frac{\partial^2}{\partial x \partial y} \right)^\alpha u \right| \le Ay^{-\beta - |\alpha|}$$

for any multi-index α.

Now we will see how the preceding theorem follows from these two lemmas. One direction is clear, for $F \in H^p$, $F = (u_0, u_1, \ldots, u_n)$ implies that $u^* \le F^* \in L^p$ and $\|u^*\|_{L^p} \le A_p \|F\|_{H^p}$.

Conversely, if u is harmonic, bounded near ∞, and $u^* \in L^p$, $p > (n-1)/n$, then by the second lemma, $|u(x, y)| \le Ay^{-n/p}$ and $|\partial u/\partial x_k| = O(y^{-n/p-1})$. If we let

$$u_j(x, y) = -\int_0^\infty \frac{\partial u}{\partial x_j}(x, y+s) \, ds,$$

then it is easy to check that $(u, u_1, \ldots, u_n)$ is a Cauchy–Riemann system which we call F. Now clearly $|u_j(x, y)| \le Ay^{-n/p}$ and, since $u^* \in L^p$, we get, by the area theorem, that $S(u) \in L^p$. Let

$$S_j^2(x) = \iint\limits_{\Gamma_\beta(x)} \left| \frac{\partial u}{\partial x_j} \right|^2 y^{1-n} \, dx dy.$$

Then $S_j \le S$ implies $S_j(u) \in L^p$. Thus if we let

$$S_y^2(u_j)(x) = \iint\limits_{\Gamma_\beta(x)} \left| \frac{\partial u_j}{\partial y} \right|^2 y^{1-n} \, dx dy,$$

then $S_y(u_j) = S_j(u) \in L^p$. But, by the first lemma, it then follows that if we define

$$S^2(u_j) = \iint\limits_{\Gamma_\alpha(x)} |\nabla u_j|^2 y^{1-n} \, dx dy,$$

then $S(u_j) \in L^p$ and $\|S(u_j)\|_p \le A_p \|u^*\|_{L^p}$. By the area theorem, therefore, $\|u_j^*\|_p \le A_p \|u^*\|_p$ so that $F \in H^p$ and $\|F\|_{H^p} \le A_p \|u^*\|_p$.

We have already proved the first theorem, so now we prove the second.
We will use the mean value property. We know that

$$u(x, y) = \left[\int_{B((x,y),y/2)} u(s, t) \, ds dt \right] \cdot \frac{1}{m(B((x, y), y/2))}.$$

By Jensen's inequality,

$$|u(x, y)|^p \le \frac{1}{m(B)} \int_B |u(s, t)|^p \, ds dt$$

$$\le \frac{C}{y^{n+1}} \int_{y/2 \le t \le 3y/2} |u(s, t)|^p \, ds dt$$

$$\le C' y^{-n}$$

or

$$|u(x, y)| \le C'' y^{-n/p}.$$

By a similar argument, if $|u(x, y)| \le C y^{-\alpha}$, and D is a differential monomial of order k, φ a real bump function, $\varphi \ge 0$, $\operatorname{supp} \varphi \subseteq B(0, 1/2)$, and $\varphi_\delta(x) = \delta^{-n-1} \varphi(x/\delta)$, then write

$$u(x, y) = \varphi_y * u(x, y).$$

Differentiation under the integral sign yields the desired result.   □

Now we come to a very tricky lemma of Hardy and Littlewood.

Lemma 5.6.8 *Fix* $0 < p < \infty$. *If* u *is harmonic on a ball* $B(0, r) \subseteq \mathbb{R}^n$, *and if* $u \in C(\overline{B})$, *then there exists* $C_p > 0$ *so that*

$$|u(0)|^p \leq \frac{C_p}{m(B)} \int_B |u(x)|^p \, dx \, .$$

Proof Assume that B is the unit ball in $\mathbb{R}^n$. Further assume that $p < 1$, otherwise the assertion is trivial by convexity (with $C_p = 1$). Finally, assume that

$$\frac{1}{m(B)} \int_B |u|^p = 1 \, .$$

For $0 < r < 1$, denote by

$$m_p(r) = \left(\int_{S^{n-1}} |u(r\xi)|^p \, d\sigma(\xi) \right)^{1/p} \, , \quad 0 < p < \infty \, .$$

Also set

$$m_\infty(r) = \sup_{\xi \in S^{n-1}} |u(r\xi)| \, .$$

We may suppose that $m_\infty(r) \geq 1$, all $0 \leq r \leq 1$, since in case $m_\infty(r) < 1$ for some r then $|u(0)| \leq 1$ so

$$|u(0)|^p \leq 1 = \frac{1}{m(B)} \int_B |u|^p \, .$$

Claim (a) There exists a θ, $0 < \theta < 1$, such that

$$m_1(r) \leq (m_p(r))^{1-\theta} \cdot (m_\infty(r))^\theta \, .$$

Proof of the Claim We have

$$\int_{S^{n-1}} |u(r\xi)| \, d\sigma(\xi) = \int_{S^{n-1}} |u|^p |u|^{1-p} \, d\sigma(\xi)$$

$$\leq \int_{S^{n-1}} |u|^p \, d\sigma \cdot \sup_{\xi \in S^{n-1}} |u(r\xi)|^{1-p}$$

$$= (m_p(r))^p \cdot (m_\infty(r))^{1-p} \, .$$

So $\theta = 1 - p$ will do.

Let us restrict attention to $1/2 < r \leq 1$. Let $0 < \rho < r$. Then

$$m_\infty(p) = \sup_{\xi \in S^{n-1}} |u(\psi)| \leq \|P_\rho^r(\xi)\|_{L^\infty} \cdot \|m_1(r)\|_{L^1},$$

where P^r is the Poisson kernel for $B(0, r)$, which is

$$\leq C(1 - \rho/r)^{-n+1} \cdot m_1(r).$$

We now let $\rho = r^a$, $a > 1$, where the constant a will be chosen in a moment. So we have

$$m_\infty(r^a) \leq C(1 - r^{a-1})^{-n+1} \cdot m_1(r).$$

Claim (b) We have

$$\int_{1/2}^1 \log m_\infty(r^a) \frac{dr}{r} \leq C_{a,\theta} + (1 - \theta) \int_{1/2}^1 \log m_p(r) \frac{dr}{r} + \theta \int_{1/2}^1 \log m_\infty(r) \frac{dr}{r}.$$

Proof of Claim (b) We see that

$$\int_{1/2}^1 \log m_\infty(r^a) \frac{dr}{r} \leq \int_{1/2}^1 \log C + \left(\log \left| 1 - \frac{r^a}{r} \right| \right) \cdot |n - 1| + \log m_1(r) \frac{dr}{r}$$

$$\leq \log C \cdot \log 2 + C' + \int_{1/2}^1 \log m_1(r) \frac{dr}{r}$$

$$\leq C'' + \int_{1/2}^1 (1 - \theta) \log m_p(r) \frac{dr}{r} + \int_{1/2}^1 \theta \log m_\infty(r) \frac{dr}{r}.$$

This is our new claim.

Thus

$$\int_{1/2}^1 \log m_\infty(r^a) \frac{dr}{r} \leq C''' + \int_{1/2}^1 \theta \log m_\infty(r) \frac{dr}{r} \qquad (\star)$$

since

$$(1 - \theta) \int_{1/2}^1 \log m_p(r) \frac{dr}{r} \leq K \int_{1/2}^1 (1 + m_p(r)) \frac{dr}{r} \leq K + K' \int_B |u|^p \leq K''.$$

By a change of variables in $(\star)$,

$$\frac{1}{a} \int_{1/2}^1 \log m_\infty(r) \frac{dr}{r} \leq C''' + \theta \int_{1/2}^1 \log m_\infty(r) \frac{dr}{r}.$$

Now we merely choose $a > 1$ so that $1/a > \theta = 1 - p$. Note that $\log m_\infty(r) \geq 0$ since $m_\infty(r) \geq 1$ by assumption. So

$$\int_{1/2}^1 \log m_\infty(r) \frac{dr}{r} \leq C''''.$$

Thus, for some r,

$$\log m_\infty(r) \leq C''''$$

or

$$m_\infty(r) \leq C^* \equiv \exp(C'''').$$

But then

$$|u(0)|^p \leq (C^*)^p = (C^*)^p \cdot \frac{1}{m(B)} \int_0^B |u(x)|^p \, dx.$$

This completes the proof of the Hardy–Littlewood lemma. $\square$

Now we make a sequence of remarks about u^*.

(1) Let $1 < p < \infty$ and u harmonic on $\mathbb{R}_+^{n+1}$. Then $u^* \in L^p$ if and only if $\sup_{y>0} \int_{\mathbb{R}^n} |u(x, y)|^p \, dx < \infty$.

(2) Let $(n + 1)/n < p < \infty$ and u harmonic. Then $u^* \in L^p$ if and only if there exists $(u_0, u_1, \ldots, u_n)$ a Cauchy–Riemann system with $u = u_0$ such that

$$\sup_{y>0} \int \left(\sum_{j=0}^n |u_j(x, y)|^2 \right)^{p/2} dx < \infty.$$

(3) Let us push this one step further, to $(n - 1)/(n + 1) < p < \infty$. Suppose that u is harmonic on $\mathbb{R}_+^{n+1}$. Then $u^* \in L^p$ if and only if there exist $u_{ij}(x, y)$, $0 \leq i, j \leq n$, such that $u(x) = u_{00}$, $u_{ij} = u_{ji}$, and $\sum_i u_{ii} = 0$. Moreover, if we define

$$u_{ijk} = \frac{\partial}{\partial x_k} u_{ij},$$

then we require that the u_{ijk} be symmetric in i, j, k and have vanishing trace. (these are the substitute for the Cauchy–Riemann equations.) We also require that

$$\int_{\mathbb{R}^n} \left(\sum_{i,j,k} |u_{ijk}(x)|^2 \right)^{p/2} dx < \infty.$$

(4) There is an analogous definition, using tensors of rank k, for the case $p > (n-1)/(n-1+k)$. Note that the case $k = 1$ coincides with **(2)**.

(4') Recall that, in the classical case, the square of the Hilbert transform is $-I$ so polynomials in the Hilbert transform are quite simple. This is not the case on $\mathbb{R}^{n+1}_+$. So a Cauchy–Riemann system is more complicated.

Now define, for $u \in h(\mathbb{R}^{n+1}_+)$, $u^+(x) = \sup_{y>0} |u(x, y)|$. Clearly

$$\|u^+\|_{L^p} \le \|u^*\|_{L^p}.$$

However, we will prove that

$$\|u^*\|_{L^p} \le C_p \|u^+\|_{L^p}, \quad 0 < p < \infty.$$

There is an example due to Bagemihl and Seidel which distinguishes radial and nontangential convergence. In fact they show that, given $f \in C(\overline{D})$ and $\mathcal{H} \subseteq S$ of first category, with $S = \partial D$, there exists an $F \in H(D)$ such that $|F - f| \to 0$ along all radii through $\mathcal{H}$.

The case for nontangential convergence is quite different. In fact, if $F \in H(D)$ and F has nontangential limit 0 on a set of positive measure, then $F \equiv 0$.

Now we will prove that

$$C_p [M(u^+(x))^{p/2}]^{2/p} \ge u^*(x).$$

In fact, with Γ fixed as usual, we have, for $(z, y) \in \Gamma(x)$, that

$$|u(x, y)|^{p/2} \le C_p \frac{1}{m(B((x, y), y/2))} \int_{B((x,y),y/2)} |u|^{p/2} dx' dy'$$

by the Hardy–Littlewood lemma. And this last is

$$\le C_p \frac{1}{m(B)} \int_B |u^+(x')|^{p/2} dx' dy'$$

$$\le C_p' \frac{1}{y^{n+1}} \cdot y \cdot \int_{|x'-x|<y/2} |u^+(x')|^{p/2} dx'$$

$$\leq C_p'' \frac{1}{m(B(x, y/2))} \int_{B(x, y/2)} |u^+(x')|^{p/2}.dx'$$

$$\leq C_p'' M(u^+(x))^{p/2}.$$

Clearly, allowing (x, y) to range over $\{|x - \overline{x}| < y\}$ will only change these estimates by a factor. So we are done. By the maximal theorem for L^2, we have

$$\||u^*|^{p/2}\|_{L^2}^2 = \|u^*\|_{L^p}^p$$

$$\leq C_p'' \|M(u^+)^{p/2}\|_{L^2}^2$$

$$\leq C_p''' \|(u^+)^{p/2}\|_{L^2}^2$$

$$= C_p''' \|u^+\|_{L^p}^p.$$

Thus

$$\|u^*\|_{L^p} \leq C_p \|u^+\|_{L^p}$$

as desired. $\square$

Lemma 5.6.9 *Suppose that $u(x, y)$ is harmonic on $\mathbb{R}_+^{n+1}$ and that $\sup_{y>0} \int_{\mathbb{R}^n} |u(x, y)|^p dx < \infty$, $0 < p < \infty$. Then $\lim_{y\to 0} u(x, y) = f$ as a tempered distribution and u is uniquely determined by f. That is, there exists a unique f a tempered distribution so that*

$$\lim_{y\to 0} \int u(x, y)\varphi(x)\, dx = f(\varphi) \quad \text{for all } \varphi \in \mathcal{S}.$$

Proof The lemma is well known for $p \geq 1$. For $p > 1$, the convergence is in L^p and for $p = 1$ the convergence is in the weak-$*$ topology of finite measures. We restrict attention to $p \leq 1$.

Let $u_\delta(x, t) = u(x, t+\delta)$, $\delta > 0$, for $(x, t) \in \mathbb{R}_+^{n+1}$. By the second lemma above, we know that

$$\|u_\delta(x, t)\|_\infty \leq A \cdot \delta^{-n/p}.$$

This, combined with the pth power integrability hypothesis, implies that

$$\int_{\mathbb{R}^n} |u_\delta(x, t)|\, dx \leq C_\delta \quad \text{for all } \delta,$$

with C_δ independent of t. Thus u_δ is the Poisson integral of a finite measure μ_δ, with μ_δ the weak-$*$ limit of $u_\delta(x, t)$ as $t \to 0$. It follows by continuity of u that

$$\mu_\delta = u_\delta(x, 0) = u(x, \delta).$$

Using the Fourier transform and corresponding identities for the Poisson kernel, we have

$$\widehat{u_\delta}(\xi, t) = \widehat{u_\delta}(\xi, 0) \cdot e^{-2\pi|\xi|t}.$$

That is to say,

$$\widehat{u}(\xi, t + \delta) = \widehat{u_\delta}(\xi, 0) \cdot e^{-2\pi|\xi|t} \equiv \widehat{u_0}(\xi)e^{-2\pi|\xi|(t+\delta)}. \qquad (*)$$

Trivially this definition of $\widehat{u_0}(\xi)$ is unambiguous in δ. Note that $\widehat{u_0}$ is a continuous function. Moreover, we know that

$$\left|\widehat{u_0}(\xi)e^{-2\pi|\xi|t}\right| \leq \int_{\mathbb{R}^n} |u(x, t)|\, dx$$

$$\leq \int_{\mathbb{R}^n} |u(x, t)|^p \cdot t^{-(n/p)\cdot(1-p)}\, dx$$

$$\leq C \cdot t^{n(1-1/p)}.$$

In conclusion,

$$|\widehat{u_0}(\xi)| \leq C \cdot t^{n(1-1/p)} \cdot e^{2\pi|\xi|t} \quad \text{for all } t > 0.$$

Letting $t = 1/|\xi|$ gives

$$|\widehat{u_0}(\xi)| \leq C|\xi|^{n(1/p-1)}.$$

Hence $\widehat{u_0}$ is of polynomial growth. Let $f = (\widehat{u_0})^\vee$ in the sense of tempered distributions. By $(*)$, for any $\varphi \in \mathcal{S}$,

$$\int_{\mathbb{R}^n} u(x, t)\overline{\varphi}(x)\, dx = \int_{\mathbb{R}^n} \widehat{u_0}(\xi)e^{-2|\xi|\pi t}\widehat{\varphi}(-\xi)\, d\xi$$

$$= f\left[\left(e^{-2\pi|\xi|t} \cdot \widehat{\varphi}(-\xi)\right)^\vee\right].$$

Since $(\epsilon^{-2\pi|\xi|t}\widehat{\varphi})^\vee \to \varphi(\xi)$ uniformly on compact sets as $t \to 0$, we obtain

$$\int_{\mathbb{R}^n} u(x, t)\overline{\varphi}(x, t)\, dx \to f(\varphi)$$

as desired. $\qquad\qquad\qquad\qquad\qquad\qquad\qquad\qquad\qquad\qquad\qquad\qquad\square$

Question Which tempered distributions arise as boundary values of H^p functions? The answer is:

Theorem 5.6.10 *Suppose that f is a tempered distribution on $\mathbb{R}^n$. Then the following are equivalent for $0 < p < \infty$:*

(1) *$f = \lim_{y \to 0} u(x, y)$ in the distribution sense, where u is harmonic and $u^* \in L^p$.*

(2) *For all $\varphi \in \mathcal{S}$ with $\varphi_\epsilon(x) = \epsilon^{-n} \varphi(x/\epsilon)$, we have $\sup_{\epsilon > 0}(f * \varphi_\epsilon)(x) \in L^p$.*

(3) *There exists a $\Phi \in \mathcal{S}$ with $\Phi_\epsilon(x) = \epsilon^{-n} \Phi(x/\epsilon)$, $\int \Phi\, dx \neq 0$, such that $\sup_{\epsilon > 0}(f * \Phi_\epsilon) \in L^p$.*

The proof of this theorem is going to take some doing, so we begin with some remarks and discussion. We ask the reader to bear in mind that the next fifteen or more pages will be dealing with this theorem.

Remark 5.6.11 Recall that a set $\mathcal{B} \subseteq \mathcal{S}$ is bounded if

$$\mathcal{B} \subseteq \{\varphi \in \mathcal{S};\ \|\varphi\|_{\alpha, \beta} \leq C \text{ for some } C \text{ and for finitely many } \alpha, \beta\}.$$

Now part (2) of the theorem can be given a more precise statement. Fix $0 < p < \infty$. Then $f \in H^p$ if and only if there exists a bounded set $\mathcal{B} \subseteq \mathcal{S}$ such that

$$\sup_{\varphi \in \mathcal{B}} \sup_{\epsilon > 0} |f * \varphi_\epsilon| \in L^p.$$

We claim that the size of $\mathcal{B}$ required, that is the amount of control on the smoothness of elements of $\mathcal{B}$, is

$$\text{(order of differentiability)} > n \cdot \left(\frac{1}{p} - 1\right),\ 0 < p \leq 1.$$

This will be seen. This constant has already come up in an earlier lemma.

Example 5.6.12

(1) The Poisson kernel for $\mathbb{R}^{n+1}_+$ is

$$\frac{c}{(1 + |x|^2)^{(n+1)/2}}.$$

This is not a Schwartz function, but we know from the previous pages that this kernel will work.

(2) The characteristic function of the unit ball will work for $p \geq 1$, even though it has no smoothness. Note that in this case the corresponding maximal function is

$$\sup_{\epsilon > 0} \left| \frac{1}{\epsilon^n} \int_{|y| \le \epsilon} f(x - y) \, dy \right| .$$

This is dominated by the Hardy–Littlewood maximal function. So it is bounded on $L^p = H^p$, $p > 1$.

Will this work for $p \le 1$? For $p < 1$, no, since the boundary value of $u \in H^p$, $p < 1$, is, in general, a distribution and we cannot talk about the absolute value of a distribution.

What about $p = 1$? Well, the class of L^1 functions such that

$$\sup_{\epsilon > 0} \frac{1}{\epsilon^n} \int_{|y| \le \epsilon} |f(x - y)| \, dy \in L^1$$

is locally the same as $L \log L$, which is not the same as H^1.

These questions have close ties with the theory of singular integral operators and pseudo-differential operators. See [FES] and also [STE1].

Consider the Riesz transform

$$(R_j f)\widehat{}(\xi) = \frac{i\xi}{|\xi|} \widehat{f}(\xi) .$$

It is the case that $f \in H^p$ implies $R_j f \in H^p$, $j = 1, \ldots, n$. Thus $Pf \in H^p$ for any polynomial P in the R_j. This is clear for $p > 1$, will follow from the duality of H^1 and BMO for $p = 1$, and follows from characterizations of the space H^p for $p < 1$ in terms of maximal operators. Recall that the H^p are defined in terms of the Riesz transforms.

Now we prove the theorem. We show that, if we know (2) for all testing functions, then we know (2) for the Poisson kernel. One can do this by piecing together P by weighted Schwartz functions as in the Fefferman–Stein paper. But there is a simple trick which gives the result.

Let us assume (2). Choose the particular Schwartz function $\varphi(x) = e^{-\pi|x|^2}$. Then $\widehat{\varphi}(\xi) = e^{-\pi|\xi|^2}$. Now

$$\widehat{P}_y(\xi) = e^{-2\pi y|\xi|}$$

and, just as

$$(P_y * f)\widehat{}(\xi) = e^{-2\pi|\xi|y} \widehat{f}(\xi) , \qquad (*)$$

we also have

$$(\varphi_{4\pi y} * f)\widehat{}(\xi) = e^{-4\pi^2 y|\xi|^2} \widehat{f}(\xi) .$$

Given a locally integrable f, if we let $W_y f$ denote $\varphi_{4\pi y} * f$, then

$$\frac{\partial W_y f(x)}{\partial y} = \sum_{j=1}^{n} \frac{\partial^2 W_y f(x)}{\partial x_j^2}.$$

This is the heat equation and $\varphi_{4\pi y}$ is the heat kernel.

Now, assuming (2), we have $\sup_{y>0} |W_y f| \in L^p$. Now we claim that

$$P_y(x) = \int_0^\infty K(u)\varphi_{\pi y^2/u}(x)\,du$$

for some function $K \geq 0$ with $\int_0^\infty K(u)\,du = 1$. We will use the classical formula

$$e^{-\alpha} = \frac{1}{\sqrt{\pi}} \int_0^\infty \frac{e^{-u}}{\sqrt{u}} e^{-\alpha^2/(4u)}\,du$$

(see [STW1]). With $\alpha = 2\pi|\xi|y$, this says

$$e^{-2\pi|\xi|y} = \frac{1}{\sqrt{\pi}} \int_0^\infty \frac{e^{-u}}{\sqrt{u}} e^{-\pi^2|\xi|^2y^2/u}\,du = \int_0^\infty \left(\frac{1}{\sqrt{\pi}}\frac{e^{-u}}{\sqrt{u}}\right) \widehat{\varphi_{\pi y^2/u}}(\xi)\,du,$$

that is to say,

$$P_y(x) = \int_0^\infty K(u)\varphi_{\pi y^2/u}(x)\,du$$

with

$$K(u) = \frac{1}{\sqrt{\pi}} \frac{e^{-u}}{\sqrt{u}}.$$

From this identity it follows easily that

$$\sup_{y>0} |f * P_y(x)| \leq \int K(u) \sup_{t>0} |\varphi_t * f(x)|\,du$$

$$\leq \sup_{t>0} |\varphi_t * f(x)| \in L^p.$$

Thus, if $u = P_y * f$, then we have $u^+ \in L^p$ so that $u^* \in L^p$ and therefore $u \in H^p$.

Thus we have learned that the theory of H^p in terms of boundary values of the heat equation instead of Laplace's equation is no different from the classical theory. Sometimes the heat kernel is nicer since the product of one-dimensional kernels gives the n-dimensional kernel. Now we show that $(1) \Rightarrow (2)$. We remark that the Fefferman–Stein paper has a mistake in this segment of the proof.

Definition 5.6.13 Suppose that $u(x, y)$ is defined on $\mathbb{R}^{n+1}_+$ and continuous. Let

$$u^+(x) = \sup_{y>0} |u(x, y)|,$$

$$u^*(x) = \sup_{|z|<y} |u(x - z, y)|,$$

$$u^*_N(x) = \sup_{|z|<Ny} |u(x - z, y)|.$$

Finally, we have an enormous maximal function

$$u^{**}_r(x) = \sup_{z,y} |u(x - z, y)| \cdot \left(\frac{y}{y + |z|}\right)^r \quad \text{for } r > 0.$$

Lemma 5.6.14 *Suppose that $u^* \in L^p$. Then $u^*_N \in L^p$ for all N and*

(a)

$$\int_{\mathbb{R}^n} |u^*_N(x)|^p \, dx \leq C_p N^n \int |u^*(x)|^p \, dx.$$

Also, if $r > n/p$, then $u^{**}_r \in L^p$.

(b)

$$\int (u^{**}_r(x))^p \, dx \leq C_{r,p} \int_{\mathbb{R}^n} (u^*(x))^p \, dx.$$

Proof Let $E_\alpha = \{x : |u^*(x)| > \alpha\}$. Then, of course,

$$\int_{\mathbb{R}^n} (u^*(x))^p \, dx = p \int_0^\infty \alpha^{p-1} m(E_\alpha) \, d\alpha. \qquad (\star)$$

Let $E^*_\alpha = \{x : u^*_N(x) > \alpha\}$. We wish to see that

$$m(E^*_\alpha) \leq C N^n m(E_\alpha).$$

Let χ_{E_α} be the characteristic function of E_α. We claim that, for C small enough,

$$E^*_\alpha \subseteq \left\{ M(\chi_{E_\alpha})(x) > \frac{C}{N^n} \right\}.$$

Thus we will have

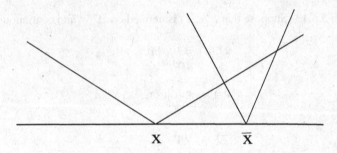

Fig. 5.9 Comparison of x and $\overline{x}$

$$m(E_\alpha^*) \le m\left\{ M(\chi_{E_\alpha}(x)) > \frac{C}{N^n} \right\} \le \frac{m(E_\alpha) \cdot N^n}{C}.$$

Then $(\star)$ will imply **(a)** of the lemma. So we prove the claim. In fact we prove the contrapositive:

$$M(\chi_{E_\alpha})(x) \le N^{-n} \Rightarrow u_N^*(x) \le \alpha.$$

See Fig. 5.9.

For C we choose

$$C = \frac{1}{2}\left(\frac{N}{N+1} \right)^n.$$

We claim now that, for each z with $|z - x| < Ny$, $y > 0$, there exists an $\overline{x}$ with $|\overline{x} - z| < y$ and $u^*(x) \le \alpha$. For if not then there exists a z' with $|x - z'| < Ny$ and $u^*(\overline{x}) > \alpha$ for all $\overline{x}$ with $|\overline{x} - z'| < y$. This means that, if we look at $B(x, (N+1)y)$, then

$$M(\chi_{E_\alpha})(x) \ge \frac{m[B(x, (N+1)y) \cap E_\alpha]}{m[B(x, (N+1)y)]}$$

$$\ge \frac{m[B(z', y)]}{V_n(N+1)^n y^n}$$

$$= \frac{V_n y^n}{V_n(N+1)^n y^n}$$

$$= \frac{1}{(N+1)^n}$$

$$= 2CN^{-n}$$

$$> CN^{-n},$$

which is a contradiction. This prove the claim.

Now if $u_N^*(x) > \alpha$, then there exist (z, y) satisfying $|x-z| < Ny$ and $|u(z, y)| > \alpha$. Therefore, for all $\bar{x}$ satisfying $|\bar{x} - z| < y$ we have $u^*(\bar{x}) > \alpha$. This contradicts the claim which we have just proved. So $u_N^*(x) \leq \alpha$ as desired.

Now part **(b)** of the lemma follows from part **(a)**. Namely,

$$\sup_{z,y} |u(x - z, y)| \cdot \left(\frac{y}{y + |z|}\right)^r \leq \sup_{0 \leq k < \infty} \left\{ \sup_{2^k y \leq |z| \leq 2^{k+1} y} |u(x - z, y)| \cdot \left(\frac{y}{y + |z|}\right)^r \right.$$

$$\left. + \sup_{|z| \leq y} |u(x - z, y)| \cdot \left(\frac{y}{y + |z|}\right)^r \right\}$$

$$\leq 2 \sup_{0 \leq k < \infty} u_{2^k}^*(x) \cdot 2^{-kr} .$$

Thus we have

$$u_r^{**}(x) \leq 2 \sup_{0 \leq k < \infty} 2^{-kr} u_{2^k}^*(x) .$$

So

$$(u_r^{**})^p \leq 2^p \sum_{k=0}^{\infty} (u_{2^k}^*(x))^p 2^{-pkr}$$

hence

$$\int (u_r^{**})^p \, dx \leq 2^p C \left(\sum_{k=0}^{\infty} 2^{kn} 2^{-pkr} \right) \int_{\mathbb{R}^n} (u^*(x))^p \, dx .$$

Now $\sum_{k=0}^{\infty} 2^{(n-pr)k}$ converges provided that $r > n/p$ as claimed. $\qquad \square$

We continue now with the proof of **(1)** $\Rightarrow$ **(2)**. The idea is to express any admissible φ as an integral of Poisson kernels. For example (*formally*), suppose that $\varphi \in S$ and f is a tempered distribution. Let $\varphi_1 = (-\triangle)^\ell \varphi$, where ℓ is an integer, and let $k = 2\ell$. We claim that

$$f * \varphi = \frac{1}{\Gamma(k)} \int u(x - z, y)\varphi_1(z)y^{k-1} \, dz dy . \qquad (\star)$$

Here we think of the function u as the Poisson integral of f. This is equivalent to

$$\widehat{f}(\xi)\widehat{\varphi}(\xi) = \frac{1}{\Gamma(k)} \int_0^\infty \widehat{f}(\xi)e^{-2\pi y|\xi|}\widehat{\varphi_1}(\xi)y^{k-1} \, dy . \qquad (*)$$

(recall that we know that $\widehat{f}$ is a continuous function of polynomial growth.) Now since

$$\int_0^\infty e^{-2\pi y|\xi|} y^k \frac{dy}{y} = \left(\frac{1}{2\pi|\xi|}\right)^k \Gamma(k),$$

the right side of $(*)$ is

$$\widehat{f}(\xi) \left(\frac{1}{2\pi|\xi|}\right)^k \widehat{\varphi_1}(\xi).$$

But $\widehat{\varphi_1}(\xi) = (2\pi|\xi|)^k \widehat{\varphi}(\xi)$, so this last expression is $\widehat{f}(\xi)\widehat{\varphi}(\xi)$, which is the left side of $(*)$. Thus we have "demonstrated" $(\star)$, which enables us to see that $f * \varphi$ can be thought of as an integral of Poisson integrals. Making all this precisely requires a considerable amount of work. To wit:

Lemma 5.6.15 *Let $\varphi \in S$. Suppose that K is a given even integer and N is an integer. Both are large and positive. We claim that there exist φ_1, φ_2 such that*

$$|\varphi_j(x)| \le A(1 + |x|)^{-N} \ , j = 1, 2,$$

and a function $\psi(y)$ continuous on $[0, 1]$ such that

$$\psi(y) = O(y^{K-1}) \text{ as } y \to 0$$

and

$$f * \varphi = \int_{\mathbb{R}^n} \int_0^1 u(x - z, y)\varphi_1(z)\psi(y)\, dy dz + \int_{\mathbb{R}^n} u(x - z, 1)\varphi_2(z)\, dz. \quad (5.1)$$

The motivation for this mess is the following: Suppose this last lemma is true. Then we have

$$|u(x - z, y)| \le u_r^{**}(x) \cdot \left(\frac{y + |z|}{y}\right)^r$$

for any y, z, r. Replace u in formula (5.1) for $f * \varphi$ by this estimate, choosing $k > r$, and we find that

$$f * \varphi(x) \le A u_r^{**}(x),$$

where K and N are chosen sufficiently large to guarantee integrability. By dilation invariance, it follows that

$$f * \varphi_\epsilon(x) \le A u_r^{**}(x).$$

But this and the lemma a few pages ago yield (2). So now we need only to prove the last lemma. We begin with a sublemma:

Sublemma *Given K and N_1, there exists a $\psi(y)$ continuous on $[0, 1]$ such that*

(1) $\frac{1}{\Gamma(k)} \int_0^1 \psi(y)e^{-2\pi|\xi|y}\,dy - \frac{1}{(2\pi|\xi|)^k} = O(e^{-c|\xi|})$, *and we want this estimate to be stable under differentiation.*

(2) $\frac{1}{\Gamma(k)} \int_0^1 \psi(y)e^{-2\pi|\xi|y}\,dy - C = O(|\xi|^{N_1})$ *as $\xi \to 0$, and we want this integral to be of class C^{N_1-1} near 0.*

Here C is a constant and K, N are fixed, positive integers.

Proof of the Sublemma We let

$$\psi_0(y) = \begin{cases} 1 & \text{if } 0 \leq y \leq 1/2 \\ \text{to be specified if} & y > 1/2. \end{cases}$$

For $y > 1/2$ we mainly want ψ_0 to be smooth, bounded, and we want the moments of $\psi_0 y^{k-1}$ up to order N_1 to be 0. That is to say, we want

$$\int_0^1 \psi_0(y)y^{j+k-1}\,dy = 0 \ , \ 1 \leq j \leq N_1 .$$

Then for $|\xi| > 1$ we have, with $\psi(y) = y^{k-1}\psi_0(y)$, that

$$\int_0^1 \psi(y)e^{-2\pi|\xi|y}\,dy = \int_0^{1/2} y^{k-1}e^{-2\pi|\xi|y}\,dy + \int_{1/2}^1 \psi_0(y)e^{-2\pi|\xi|y}\,dy \equiv T_1 + T_2 .$$

Now

$$|T_2| \leq C \cdot \frac{1}{2} \sup_{y\in[1/2,1]} e^{-2\pi|\xi|y} \leq C'e^{-\pi|\xi|} .$$

Integration of T_1 by parts k times gives

$$\left| T_1 - \frac{1}{|2\pi\xi|^k} \right| \leq Ce^{-|\xi|} .$$

This last estimate is stable under differentiation, as we see by writing T_1 as

$$|\xi|^{-k} \int_0^{|\xi|/2} y^{k-1}e^{-2\pi y}\,dy ,$$

differentiating and repeating the estimates. The estimate for $|\xi|$ small follows from writing the Taylor series for $e^{-2\pi|\xi|y}$ and using the fact that the moments up to N_1

of ψ vanish. To see how to write a function with vanishing moments, consult Stein, *Singular Integrals*.

Equipped with the sublemma, let us now prove the lemma. Denote

$$M(\xi) = \int_0^1 \psi(y) e^{-2\pi|\xi|y}\, dy\,,$$

where ψ is given by the sublemma with $N_1 = N + 1$. The desired equality is equivalent with

$$\widehat{\varphi}(\xi) = \widehat{\varphi}_1(\xi)\left(\int_0^1 e^{-2\pi|\xi|y}\psi(y)\,dy\right) + e^{-2\pi|\xi|}\widehat{\varphi}_2(\xi)$$

$$= \widehat{\varphi}_1(\xi)M(\xi) + e^{-2\pi|\xi|}\widehat{\varphi}_2(\xi)\,.$$

This is what we will show, for appropriate φ_1, φ_2.

Now φ is given and we write

$$\widehat{\varphi} = \widehat{\varphi}_3 + \widehat{\varphi}_4\,,$$

where $\varphi_j \in \mathcal{S}$, $\widehat{\varphi}_3$ is supported either near 0 or outside a large ball, and the support of $\widehat{\varphi}_4$ is a compact neighborhood of 0.

Choose φ_2 such that

$$\widehat{\varphi}_2 e^{-2\pi|\xi|} = \widehat{\varphi}_4(\xi) \tag{5.2}$$

and choose φ_1 such that

$$\widehat{\varphi}_1(\xi)M(\xi) = \widehat{\varphi}_3(\xi)\,. \tag{5.3}$$

From (3), $\widehat{\varphi}_1(\xi) = M(\xi)^{-1} \cdot \widehat{\varphi}_3(\xi)$, so that $\widehat{\varphi}_1$ is C^{N-1} near 0 and is majorized by $C|\xi|^k|\widehat{\varphi}_3|$ away from 0, so we find that $|\varphi_1(x)| \le A(1 + |x|)^{-N}$ as we wish.

Now, since $\widehat{\varphi}_4$ is compactly supported and smooth, we see that $\widehat{\varphi}_2 = e^{2\pi|\xi|}\widehat{\varphi}_4$ is Schwartz, hence φ_2 is Schwartz. So $|\varphi_2(x)| \le A(1 + |x|)^{-N}$. Thus φ_1, φ_2 are constructed as specified and the lemma is proved.

Next we prove that (3) implies (2). Namely, we show that if $\sup_{\epsilon>0}|f * \Phi_\epsilon| \in L^p$ for some $\Phi \in \mathcal{S}$ with $\int \Phi dx = 1$, then $\sup_{\epsilon>0}|f * \varphi_\epsilon| \in L^p$ for all $\varphi \in \mathcal{S}$. In fact the latter supremum may be taken over appropriate bounded subsets of $\mathcal{S}$. By letting the parameters which define these subsets tend to infinity, we see that they exhaust all of $\mathcal{S}$. Our main lemma will be:

Lemma 5.6.16 *With Φ given as above, define $u(x, y) = f * \Phi_y$, $y > 0$. Then $u^* \in L^p$, where*

$$u^*(x) = \sup_{|z-x|<y} |u(z,y)|.$$

We defer the proof of this lemma and first see how the statement **(3)** $\Rightarrow$ **(2)** follows from the lemma. A construction we will make in the present argument will be used in the proof of the lemma and we beg the reader's indulgence to check that the construction is independent of the present context and the argument is not circular.

We begin with some remarks. The assumption that $\int \Phi(x)\,dx = 1$ means that $\widehat{\Phi}(0) = 1$ so $\widehat{\Phi}$ does not vanish in a neighborhood of 0. Moreover, since $\widehat{\Phi}_y(\xi) = \Phi(\xi y)$, it follows that the set where $\widehat{\Phi}_y$ does not vanish grows large as y gets small. Roughly speaking, we can, via translates and dilates of $\widehat{\Phi}$, cover any part of $\mathbb{R}^n$, hence generate any function. Now, assuming this last lemma, we have $u_r^{**} \in L^p$.

Claim 1 There exists $\mathcal{B}_1 \subseteq \mathcal{S}$, a bounded subset, such that if $\psi \in \mathcal{B}_1$ and $\varphi = \psi * \Phi_y$, then

$$|f * \varphi(x)| \le Cy^{-r} u_r^{**}(x) \, , \, 0 \le y < 1.$$

Proof of the Claim Notice that

$$f * \varphi(x) = (f * \Phi_y) * \psi(x)$$

$$= \int_{\mathbb{R}^n} u(x-z, y)\psi(z)\,dz.$$

But

$$|u(x-z, y)| \le u_r^{**}(x) \cdot \left(\frac{y}{y+|z|}\right)^{-r},$$

hence

$$|f * \varphi(x)| \le u_r^{**}(x) \int_{\mathbb{R}^n} \left(\frac{y}{y+|z|}\right)^{-r} |\psi(z)|\,dz$$

$$\le u_r^{**}(x) \cdot y^{-r} \int_{\mathbb{R}^n} (1+|z|)^r |\psi(z)|\,dz$$

$$\le Cu_r^{**}(x) \cdot y^{-r}.$$

It is clear from these estimates what the inequalities defining $\mathcal{B}_1$ should be.

Claim 2 If $\varphi \in \mathcal{B}_1$, and φ is supported in $|\xi| < 2^j$, then φ can be written as $\varphi = \psi * \Phi_y$ for some $\psi \in \mathcal{B}_1$ and $y = C \cdot 2^{-j}$. (C is of course independent of j.)

Proof of the Claim Write

$$\widehat{\psi}(\xi) = \frac{\widehat{\varphi}(\xi)}{\widehat{\Phi}(y\xi)} .$$

To insure that this makes sense, recall that $\widehat{\Phi}(0) = 1$ so that $|\widehat{\Phi}(\xi) > 1 - \epsilon$ for $\xi \in B(0, c)$, some small $c > 0$. Hence, letting $y = 2^{-j}c$ ensures that $|\widehat{\Phi}(y\xi)| > 1 - \epsilon$ on the support of $\widehat{\varphi}$. With this definition of $\widehat{\psi}$ we have $\widehat{\psi} \in S$. So $\psi \in S$ and $\varphi = \psi * \Phi_y$. Trivially, $\psi \in \mathcal{B}_1$, for $\mathcal{B}_1$ is some set of the form

$$\mathcal{B}_1 = \left\{ \varphi \in S : \sup_x (1 + |x|)^N \sum_{|\alpha| \le N} \left| \left(\frac{\partial}{\partial x} \right)^\alpha \varphi(x) \right| < N \right\},$$

where N can best be chosen, depending on n and p, at the end of the proof. Moreover, by Claim 1,

$$\sup_{|x-z|<t} |\varphi_t * f(z)| \le C \cdot 2^{jr} \cdot u_r^{**}(x),$$

where we are appealing to dilation invariance and a minor modification of the proof of Claim 1 to account for the cone condition.

Claim 3 Let $0 < K < N/4$ be a given constant. We claim that any $\varphi \in \mathcal{B}_1$ may be written in the form $\varphi = \sum_{j=0}^{\infty} C_j \varphi^{(j)}$, with $\varphi^{(j)} \in \mathcal{B}_1$ and $\widehat{\varphi^{(j)}}$ supported in $\{|\xi| < 2^j\}$ and $C^j = O(2^{-k_j})$.

Proof of the Claim Let

$$\widehat{\psi}_j = \begin{cases} 1 & \text{if} \quad 2^{j-2} < |\xi| < 2^{j-1} \\ 0 & \text{if } |\xi| > 2^j \text{ or } |\xi| < 2^{j-3} . \end{cases}$$

Define $\widehat{\psi}_j$ to be C^∞ elsewhere, $j = 1, 2, 3, \ldots$. Let $\widehat{\psi}_0 \in C_0^\infty$ be defined to be identically 1 for $|\xi| < 1/2$ and equal to 0 for $|\xi| > 1$. Define

$$\widehat{\eta}_j = \frac{2^{k_j} \cdot \widehat{\psi}_j}{\sum \widehat{\psi}_j} .$$

Then of course

$$\varphi = \sum_j 2^{-k_j} \eta_j * \varphi = \sum_j 2^{-k_j} \varphi^{(j)},$$

and the claim is proved.

Now we put the claims together: Given $\varphi \in S$ and $\mathcal{B}_1$ an appropriate bounded set containing φ, decompose φ as in Claim 3. We have

$$\sup_{|z-x|<t} |\varphi_t * f(z)| \le \sum_{j=0}^{\infty} 2^{-kj} \sup_{|z-x|<t} |\varphi_t^{(j)} * f(z)|$$

$$\le \sum_{j=0}^{\infty} C \cdot 2^{(r-k)j} u^{**}(x)$$

$$\le C \cdot u_r^{**}(x)$$

provided that we pick $k > r$. Thus we have $f^*(x) \le Cu^{**}(x)$ for all $x \in \mathbb{R}^n$ and so

$$\|f^*\|_{L^p} \le C\|u_r^{**}\|_{L^p} \le C\|u^*\|_{L^p}$$

and this is what we wished to show.

Now we need only prove the last lemma. With Φ fixed and $u(x, y) = f * \Phi_y(x)$ we have, by hypothesis, that $u^+ \in L^p$ and we wish to see that $u^* \in L^p$. To verify this, we will dominate u^* by $M(x)$, where

$$M(x) = \left(M(u^+(x)^{p/2}) \right)^{2/p}.$$

Then we will have

$$\|u^*\|_{L^p}^p \le \|M(x)\|_{L^p}^p$$

$$\le \int (M(u^+(x)^{p/2}))^2 \, dx$$

$$\le C_2 \int (u^+(x)^{p/2})^2 \, dx$$

$$= C_2\|u^+\|_{L^p}^p$$

as desired. With this goal in mind, we set

$$\Phi^j = \frac{\partial \Phi}{\partial x_j} \; , \quad U_j(x, y) = f * \Phi_y^j \; , \quad \text{and} \quad U^*(x) = \sum_{j=1}^{n} U_j^*(x).$$

Then, by the reasoning in Claims 1, 2, and 3 above, we have

$$\|U^*\|_{L^p} \le C\|u^*\|_{L^p}.$$

Now let $G = \{x : U^*(x) \le Bu^*(x)\}$, where B is a large constant to be chosen later. It suffices for us to prove the following two assertions:

(a) $\int_{\mathbb{R}^n} (u^*(x))^p \, dx \le 2 \int_G (u^*(x))^p \, dx$ for B large enough.
(b) $u^*(x) \le C \cdot M(x)$ for $x \in G$.

We first attack **(a)**. Of course

$$\int_{\mathbb{R}^n} (u^*(x))^p \, dx = \int_{\mathbb{R}^n \setminus G} (u^*(x))^p \, dx + \int_G (u^*(x))^p \, dx \, .$$

In $\mathbb{R}^n \setminus G$, $u^*(x) \leq (1/B) \cdot U^*(x)$, so this last is

$$\leq \int_{\mathbb{R}^n \setminus G} \frac{1}{B^p} \left(U^*(x) \right)^p \, dx + \int_G (u^*(x))^p \, dx$$

$$\leq \frac{C}{B^p} \int_{\mathbb{R}^n \setminus G} (u^*(x))^p \, dx + \int_G (u^*(x))^p \, dx \, .$$

If we pick B so large that $C/B^p < 1/2$, then we obtain **(a)**.

Technically, the above is not quite correct since we do not know that

$$\int_{\mathbb{R}^n \setminus G} (u^*(x))^p \, dx < \infty \, .$$

But an approximation argument legitimizes the reasoning (c.f. [FES]). One approximation that works is

$$u_{\epsilon, N}^*(x) = \sup_{|z| < y < \epsilon} |u(x - z, y)| \cdot \left(\frac{y}{\epsilon + y} \right)^N \cdot (1 + \epsilon |z|)^{-N} \, .$$

Now we prove **(b)**. We have

$$u(x, y) = f * \Phi_y(x) = \int_{\mathbb{R}^n} f(x - z) \Phi(z/y) y^{-n} \, dz \, ,$$

hence

$$y \frac{\partial u}{\partial x_j}(x, y) = U_j(x, y) \, . \tag{$\star$}$$

Let $c = 1/(8B)$. Fix $x \in G$. Choose (z, y) with $|z - x| < y$ and $|u(z, y)| > (1/2) u^*(x)$. Let $\bar{x} \in \mathbb{R}^n$ be any point satisfying $|\bar{x} - z| < y$. We will estimate $u(z, y)$ in terms of $u^+(\bar{x})$. Since the estimate will hold uniformly over $|\bar{x} - z| < cy$, we will obtain $u^*(x) \leq 2|u(z, y)| \leq KM(x)$.

Now

$$|u(z, y) - u(\bar{x}, y)| \leq |\bar{x} - z| \cdot \|\text{grad}_x(u(\,\cdot\,, y))\|_{L^\infty}$$

$$\leq cy U^*(x) \cdot \frac{1}{y}$$

$$\leq cB u^*(x)$$

$$< \frac{1}{4} u^*(x)$$

$$\leq \frac{1}{2} |u(z, y)| .$$

Therefore

$$|u(z, y)| \leq 2|u(\overline{x}, y)| \leq 2u^+(\overline{x})$$

or

$$u^*(x)^{p/2} \leq 2^p u^+(\overline{x})^{p/2} .$$

But

$$M(x) = \left(M((u^+(x))^{p/2}) \right)^{2/p}$$

$$\geq \left(\frac{m(\{|z - \overline{x}| < cy\}) \cdot u^+(\overline{x})^{p/2}}{V_n y^n} \right)^{2/p}$$

$$\geq c^n u^+(\overline{x})$$

$$\geq c^n u^*(x) .$$

This is the desired estimate. That completes the proof of the big theorem that we stated about fifteen pages ago. □

OPEN PROBLEM On $\mathbb{R}^n$, an isotropic dilation structure is given by a family of dilations

$$\alpha_\delta(x) = (\delta x_1, \delta x_2, \ldots, \delta x_n) ,$$

that is, all directions are stretched the same amount. An example of a non-isotropic dilation structure is given by the dilations

$$\alpha_\delta^*(x) = (\delta^{a_1} x_1, \delta^{a_2} x_2, \ldots, \delta^{a_n} x_n)$$

for $a_1, a_2, \ldots, a_n$ fixed and (possibly) distinct.

Thus, if $\varphi \in \mathcal{S}$, we set

$$\varphi_\delta(x) = \varphi(\alpha_{1/\delta}^*(x)) \cdot \delta^{-\sum a_j} .$$

We endeavor to formulate the above theorem in this new context.

(1') ???
(2') For all $\varphi \in \mathcal{S}$, $\sup_\delta |f * \varphi_\delta(x)| \in L^p$.

(3') For one $\Phi \in \mathcal{S}$, **(2')** holds.

And what should the last big lemma be in this new context?

For **(1')** we would want to use some analogue of the condition on f in terms of a Poisson-type integral. We also might consider some analogue of the area integral and ask that $Sf \in L^p$ (see the Fefferman–Stein paper). We also might want to consider stability under an appropriate class of non-isotropic singul.ar integral operators.

So the first problem is to prove the equivalence of all the above, appropriately formulated. There are partial results:

In an article by Calderón-Torchinsky [CAT], they show that $u^* \in L^p$ if and only if $S(u) \in L^p$, with an appropriate formulation. Everything they do is in the context of the heat equation instead of Laplace's equation. They also do not discuss arbitrary φ but instead concentrate on $e^{-|x|^2}$. So there are several directions for generalization.

Problem 2 In the Lie group context, one requires that dilations be certain automorphisms of the group. What are the analogues of the Riesz transforms in this context?

5.7 $H^1(\mathbb{R}^n)$ and Duality with Respect to *BMO*

Define

$$H^1(\mathbb{R}^n) = \{ f \in L^1 : R_j f \in L^1 \text{ for all } j \}$$

with

$$\|f\|_{H^1} \equiv \|f\|_{L^1} + \sum_{j=1}^{n} \|R_j f\|_{L^1}.$$

Theorem 5.7.1 *All of the following are equivalent:*

(1) $f \in H^1$.
(2) *There exists a u harmonic such that $u^* \in L^1$ and $\lim_{y \to 0} u(x, y) = f$ almost everywhere and in L^1.*
(3) $S(u) \in L^1$.
(4) *For $\varphi \in \mathcal{S}$, we have $\sup_y |f * \varphi_y| \in L^1$.*
(5) *There exists one $\varphi \in \mathcal{S}$ with $\int \varphi \, dx \neq 0$ such that $\sup_y |f * \varphi_y| \in L^1$.*
(6) *With $K(x) = \Omega(x)/|x|^n$, $\Omega \in C^\infty(S^{n-1})$, Ω homogeneous of degree 0, $\int_{|x|=1} \Omega(x) \, d\sigma(x) = 0$, we have $f * K \in L^1$, i.e., $\widehat{f} \cdot \widehat{K} \in \widehat{L^1}$.*

We have proved the equivalence of **(1)**, **(2)**, **(4)**, **(5)**. For the others see Fefferman–Stein, or read on.

Definition 5.7.2 Let $f \in L^1_{\text{loc}}(\mathbb{R}^n)$. Let Q be a cube in $\mathbb{R}^n$ with sides parallel to the axes. Let $f_Q = (1/|Q|) \int_Q f(t) \, dt$. This is well defined. Let

$$\|f\|_* = \sup_Q \frac{1}{|Q|} \int_Q |f - f_Q|\, dx .$$

If $\|f\|_* < \infty$, then we say that $f \in BMO$, any $\|f\|_{BMO} = \|f\|_*$.

This definition is due to John and Nirenberg [JON].

(1) The space BMO does not see constants. So a sample check to see if some notion is appropriate for BMO is to see whether it sees constants.

(2) If there exists an M such that, for all Q there exists C_Q with

$$\frac{1}{m(Q)} \int_Q |f - C_Q|\, dx \le M ,$$

then $f \in BMO$.

Proof For any Q,

$$\frac{1}{|Q|} \int_Q |f - f_Q|\, dx \le \frac{1}{|Q|} \left[\int_Q |f - C_Q|\, dx + \int_Q |C_Q - f_Q|\, dx \right]$$

$$\le M + \frac{1}{|Q|} \int_Q \left| C_Q - \frac{1}{|Q|} \int_Q f(t)\, dt \right| dx$$

$$\le M + \frac{1}{|Q|} \int_Q \frac{1}{|Q|} \int_Q |C_Q - f(t)|\, dt\, dx$$

$$\le M + M$$

$$= 2M .$$

So $f \in BMO$.

(3) We get the same class of functions if we use balls instead of cubes or if we use dilations of any fixed set, or cubes with any orientation.

(4) For any constant D, $\|D\|_* = 0$.

(5) If $f \in L^\infty$, then $f \in BMO$.

(6) If $f \in BMO$, then $|f| \in BMO$. This follows from

$$\big\| |f| - |f_Q| \big\| \le |f - f_Q| .$$

There do exist unbounded BMO functions. For example, $f(x) = \log |x|$. Let us first test this f on cubes centered at 0. Let Q be such a cube with side δ. For C_Q we use $\log \delta$. Then

$$\frac{1}{m(Q)} \int_Q |\log |x| - C_Q|\, dx = \frac{1}{m(Q)} \int_Q |\log |x| - \log \delta|\, dx$$

$$= \frac{1}{m(Q)} \int_Q |\log(|x|/\delta)|\,dx$$

$$= \frac{1}{m(Q^*)} \int_{Q^*} |\log|x||\,dx$$

$$= A\,,$$

where Q^* is the unit cube, by a change of variables, and this last expression is some constant A.

For a cube in general position there are two cases:

(a) $\operatorname{diam} Q \leq 2\operatorname{dist}(Q,0)$.
(b) $\operatorname{diam} Q > 2\operatorname{dist}(Q,0)$.

In case **(a)**, if x, y are any two points of Q, then $|x| \leq 2|y| \leq 4|x|$ so log is essentially constant on Q and

$$\frac{1}{|Q|} \int_Q |f - f_Q|\,dx \leq 2\log 2\,.$$

In case **(b)**, if Q^{**} is the cube with the same center as Q but sides of thrice the length, then Q^{**} contains 0. Moreover, Q^{**} is contained in a cube centered at 0 and of side not more than $10\operatorname{diam} Q$, and it contains a cube centered at 0 and of side $\operatorname{diam} Q$. It follows that

$$\frac{1}{|Q|} \int_Q |f - f_Q|\,dx \leq (120)^n \cdot A\,,$$

where A is as above. So $\log|x| \in BMO$.

Now $\log|x|$ on $\mathbb{R}^1$ has graph as shown in Fig. 5.10.

On the other hand, $\operatorname{sgn} x\, \log|x|$ is *not BMO* (Fig. 5.11).

Fig. 5.10 The graph of $\log|x|$

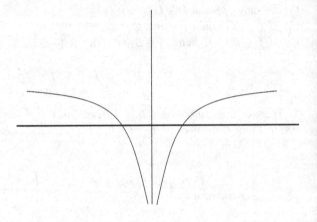

Fig. 5.11 The graph of
sgn x log $|x|$

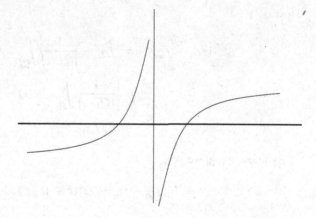

To see this, observe that (sgn x log $|x|)_Q$ for any Q centered at 0 is 0. So

$$\frac{1}{|Q|} \int_Q |f(x) - f_Q| \, dx = \frac{1}{|Q|} \int_Q |\log|x|| \, dx \to |\log 0| = \infty \,.$$

Note that *BMO* functions do not grow too fast. In fact, log $|x|$ is about as bad as they get.

FACT If f is *BMO* and Q is the unit cube at 0, then

$$\int \frac{|f(x) - f_Q|}{(1 + |x|)^{n+1}} \, dx \le C \|f\|_{BMO} \,.$$

In particular, this means that the Poisson integral of f will make sense.

Proof of the Fact Let Q be the unit cube centered at 0 and let Q_k be the cube of side 2^k centered at 0. We claim that

$$|f_Q - f_{Q_{k+1}}| \le 2^n \|f\|_* \,.$$

Well,

$$\frac{1}{|Q_k|} \int_{Q_k} |f - f_{Q_k}| \, dx \le \|f\|_* \,,$$

so that

$$\frac{1}{|Q_{k-1}|} \int_{Q_{k-1}} |f - f_{Q_k}| \, dx \le 2^k \|f\|_* \,.$$

Thus

$$|f_{Q_k} - f_{Q_{k-1}}| = \left| f_{Q_k} - \frac{1}{|Q_{k-1}|} \int_{Q_{k-1}} f(t)\, dt \right|$$

$$= \left| \frac{1}{|Q_{k-1}|} \int_{Q_{k-1}} f(t) - f_{Q_k}\, dt \right|$$

$$\leq 2^n \|f\|_* .$$

By telescoping, we get

$$|f_Q - f_{Q_k}| \leq 2^n (k+1) \|f\|_* .$$

Hence

$$\int \frac{|f(x) - f_Q|}{(1+|x|)^{n+1}}\, dx = \int_{|x| \leq 1} + \sum_{k=0}^{\infty} \int_{2^k \leq |x| \leq 2^{k+1}}$$

$$\leq \int_{|x| \leq 1} |f(x) - f_Q|\, dx + \sum_{k=0}^{\infty} 2^{-k(n+1)} \int_{|x| < 2^{k+1}} |f(x) - f_Q|\, dx$$

$$\leq 2^n \|f\|_{BMO} + \sum_{k=0}^{\infty} 2^{-k} \frac{1}{|Q_k|} \int_{|x| < 2^k} |f(x) - f_{Q_k}|\, dx$$

$$+ \sum_{k=0}^{\infty} 2^{-k} \frac{1}{|Q_k|} \int_{|x| < 2^{k+1}} |f_{Q_k} - f_Q|\, dx$$

$$\leq 2^n \|f\|_{BMO} + \sum_{k=0}^{\infty} C \cdot 2^{-k} \|f\|_{BMO} + \sum_{k=0}^{\infty} C \cdot 2^{-k} \cdot 2^n (k+1) \|f\|_{BMO}$$

$$\leq C_1 \|f\|_{BMO} .$$

FACT Standard singular integrals are bounded on BMO. More precisely,

Theorem 5.7.3 *Suppose that* $K : \mathbb{R}^n \to \mathbb{C}$ *is locally integrable away from 0. Suppose further that*

(1) $\lim_{\epsilon \to 0} \int_{|y| \geq \epsilon} k(y) f(x - y)\, dy \equiv Tf(x)$ *exists in* L^2 *norm and that* $\|Tf\|_2 \leq A \|f\|_2$.

(2) $\int_{|x| \geq 2|y|} |K(x - y) - K(x)|\, dx \leq C$ *for all* $y \neq 0$.

Then, for all $f \in L^\infty$ *with bounded support, we have* $Tf \in BMO$ *and* $\|Tf\|_{BMO} \leq CA \|f\|_{L^\infty}$.

Proof We first remark that the canonical example here is that on $\mathbb{R}^1$ the Hilbert transform of $\operatorname{sgn} x$ is $\log |x|$ which is "barely BMO."

Now let Q be any cube. By the translation invariance of everything in sight, we may suppose that Q is centered at 0. We seek a C_Q such that, if $F = Tf$, then

$$\frac{1}{|Q|} \int_Q |F(x) - C_Q| \, dx \le CA\|f\|_{L^\infty}.$$

We let $\widetilde{Q}$ denote the 3-fold dilate of Q. Note that $x \in Q$, $y \in {}^c\widetilde{Q}$ implies that $|y| \ge 2|x|$. Write

$$F = F_1 + F_2 = Tf_1 + Tf_2,$$

where $f_1 = f \cdot \chi_{\widetilde{Q}}$ and $f_2 = f \cdot \chi_{{}^c\widetilde{Q}}$. Thus

$$\|F\|_* \le \|F_1\|_* + \|F_2\|_*.$$

Now for F_1 we use the constant $C_Q = 0$. So

$$\frac{1}{m(Q)} \int_Q |F_1| \, dx \le \left(\frac{1}{m(Q)} \int_Q |F_1|^2 \, dx \right)^{1/2}$$

$$\le \left(\frac{1}{m(Q)} \int_{\mathbb{R}^n} |F_1|^2 \, dx \right)^{1/2}$$

$$\le \left(\frac{1}{m(Q)} A^2 \|f_1\|_2^2 \right)^{1/2}.$$

But f_1 is bounded and lives on $\widetilde{Q}$. So this is

$$\le CA\|f\|_{L^\infty}.$$

For F_2, we let $C_Q = \int K(-y) f_2(y) \, dy$. Note that this is finite since f_2 is bounded and supported on a compact set away from 0 and K is locally integrable away from 0. So we have

$$|F_2(x) - C_Q| = \left| \int [K(x - y) - K(-y)] f_2(y) \, dy \right|.$$

We claim that, for $x \in Q$, this expression is $\le CA\|f\|_{L^\infty}$. Assuming this claim, then we may average the inequality over Q and obtain the desired result. So let us now prove the claim: We have

$$|F_2(x) - C_Q| \le \int |K(x - y) - K(-y)| \cdot |f_2(y)| \, dy$$

$$\le \|f\|_{L^\infty} \int_{\substack{x \in Q \\ y \notin \widetilde{Q}}} |K(x - y) - K(-y)| \, dy$$

$$\leq \|f\|_{L^\infty} \int_{|y| \geq 2|x|} |K(x - y) - K(-y)| \, dy$$

$$\leq A \|f\|_{L^\infty}.$$

It appears that we have proved more than we claimed. A quick check of the proof reveals that

$$\frac{1}{m(Q)} \int_Q |F(x) - F_Q|^2 \, dx \leq M < \infty,$$

with M independent of Q. With the exponent 2 inside the integral, the norm appears stronger. In fact it is not, because of the following:

Theorem 5.7.4 (John–Nirenberg) *Suppose that* $f \in BMO$, *f supported on a fixed cube Q_0. Then*

$$m\{x \in Q_0 : |f(x) - f_{Q_0}| > \alpha\} \leq C \cdot \exp(-c_2 \alpha / \|f\|_*) \cdot m(Q_0).$$

Corollary 5.7.5 *If* $f \in BMO(Q_0)$, *then f satisfies the BMO inequality for any exponent p.*

Proof of the Corollary We have

$$\frac{1}{m(Q)} \int_Q |f - f_Q|^p \, dx = \frac{1}{m(Q)} \cdot p \cdot \int_0^\infty t^{p-1} \, dm_\alpha(t).$$

That proves the result.   □

Proof of the Theorem We first need a lemma.

Lemma 5.7.6 (Calderón–Zygmund) *Suppose that* $f \geq 0$ *is in* $L^1(\mathbb{R}^n)$ *and* $\alpha > 0$. *Then there exists closed cubes Q_k with disjoint interiors so that*

(1) $f \leq \alpha$ *almost everywhere on* $\mathbb{R}^n \setminus [\cup_k Q_k]$.
(2) $\alpha < \frac{1}{|Q_k|} \int_{Q_k} f(x) \, dx \leq 2^n \alpha$.
(3) $m(\cup_k Q_k) \leq \frac{1}{\alpha} \int_{\mathbb{R}^n} f(x) \, dx$.

Proof of the Lemma Divide $\mathbb{R}^n$ into a grid of cubes each of which is so large that the average over each one is $\leq \alpha$. Chop each of these cubes into 2^n congruent subcubes. See Fig. 5.12.

The average over each of these subcubes is $\leq 2^n \alpha$. If the average over any of these subcubes is $> \alpha$, then save it for the collection $\{Q_k\}$. On any of the remaining subcubes we have that

$$\frac{1}{|Q|} \int_Q f \, dx \leq \alpha.$$

Fig. 5.12 A cube divided
into subcubes

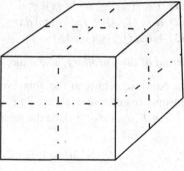

Fig. 5.13 Further chopping
into subcubes

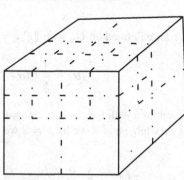

t

Chop up each of these remaining cubes into 2^n congruent subcubes (Fig. 5.13).

Repeat the process indefinitely. The situation is now as follows: If $x \notin \cup_k Q_k$, then x must be contained in arbitrarily small cubes, the average over which is $\leq \alpha$. Thus, by the Lebesgue differentiation theorem, for almost every point x of $^c(\cup_k Q_k)$, we have $f(x) \leq \alpha$. So we have proved (**1**) and (**2**). Also (**3**) follows from (**2**) since

$$|Q_k| \leq \frac{1}{\alpha} \int_{Q_k} f \, dx \,,$$

each k, and the Q_k have disjoint interiors so that

$$|\cup_k Q_k| \leq \sum_k |Q_k| = \frac{1}{\alpha} \sum_k \int_{Q_k} f \, dx \leq \frac{1}{\alpha} \int_{\mathbb{R}^n} f \, dx \,.$$

That proves the lemma. □

A corollary of the lemma is this:

Corollary 5.7.7 *If Q is a cube in $\mathbb{R}^n$, $f \in L^1(Q)$, and $\alpha > (1/|Q|) \int_Q |f(x)| \, dx$, then there exist cubes $Q_k \subseteq Q$ with disjoint interiors such that*

(1) $x \in {}^c[\cup_k Q_k] \Rightarrow |f(x)| \le \alpha$.
(2) $\alpha < (1/|Q_k|) \int_{Q_k} |f(x)| \, dx \le 2^n \alpha$.
(3) $m(\cup_k Q_k) \le (1/\alpha) \|f\|_{L^1}$.

Proof of the Corollary   The same.   □

Now we return to the John-Nirenberg inequality. It suffices to establish this inequality in the special case $\|f\|_* = 1$ and $\int_{Q_0} f(x) \, dx = 0$.

Let $F(\alpha)$, for $\alpha \ge 0$, be the least number such that

$$m\{x : |g(x)| > \alpha\} \le F(\alpha) \cdot \int_{Q_0} |g(x)| \, dx$$

for all g defined on Q_0 with $\|g\|_* = 1$, $\int g(x) \, dx = 0$. We shall prove that

$$F(\alpha) \le \frac{1}{\beta} F(\alpha - 2^n \beta) \text{ if } \beta \ge 1 \text{ and } \alpha \ge 2^n \beta.$$

Combined with the obvious estimate $F(\alpha) \le 1/\alpha$, this will yield the result. For if α is very large and if we let $\beta = 2$ we get

$$F(\alpha) \le \frac{1}{2} F(\alpha - 2 \cdot 2^n) \le \frac{1}{4} \cdot F(\alpha - 4 \cdot 2^n) \le \cdots \le C \left(\frac{1}{2}\right)^{[\alpha/2^{n+1}]} \approx C_1 e^{-C_2 \alpha}.$$

This gives the result when α is large, say $\alpha > 4^n$. But, when $\alpha \le 4^n$, we have

$$m\{x : |g(x)| > \alpha\} \le |Q_0| \le e^2 e^{-4^{-n}\alpha} m(Q_0) \le C_1 e^{-C_2 \alpha / \|f\|_*} |Q_0|.$$

So we may concentrate on proving that $F(\alpha) \le F(\alpha - 2^n \beta)$. Let us apply the Calderón–Zygmund lemma to $|f|$ at height β. We get cubes $Q_k \subseteq Q_0$ with disjoint interiors so that

$$\beta < \frac{1}{|Q_k|} \int_{Q_k} |f| \, dx \le 2^n \beta, \text{ etc.}$$

So

$$m\{x \in Q_0 : |f(x) > \alpha\} = \sum_k m\{x \in Q_k : |f(x)| > \alpha\}$$

$$\le \sum_k m\{x \in Q_k : |f(x) - f_{Q_k}| > \alpha - 2^n \beta\}.$$

But we have that $\int_{Q_k} (f(x) - f_{Q_k}) \, dx = 0$ and $\|f(x) - f_{Q_k}\|_{BMO(Q_k)} \le 1$. Therefore the last sum is

$$\le \sum_k F(\alpha - 2^n\beta) \int_{Q_k} |f(x) - f_{Q_k}| \, dx \le \sum_k F(\alpha - 2^n\beta)|Q_k| \le \frac{1}{\beta} F(\alpha - 2^n\beta) \int_{Q_0} |f| \, dx.$$

By the definition of $F(\alpha)$, it follows that

$$F(\alpha) \le \frac{1}{\beta} F(\alpha - 2^n\beta).$$

That proves John–Nirenberg. $\qquad\square$

Before we discuss duality, there are some technical points which ought to be cleared up.

Technical Point 1 If $H_0^1 = H^1 \cap S$ then H_0^1 is dense in H^1. In fact, we shall prove that H_{00}^1 is dense in H^1, where

$$H_{00}^1 = \{f \in L^2 : \widehat{f} \in C^\infty \text{ and } \widehat{f} \text{ vanishes in a}$$

$$\text{neighborhood of } 0 \text{ and near } \infty\}.$$

Since $H_{00}^1 \subseteq H_0^1 \subseteq H^1$, this will do the trick. The proof is technical so we divide it into steps.

Step α We claim that

$$\{f \in H^1 : f \text{ vanishes in a neighborhood of } 0 \text{ and in a neighborhood of } \infty\}$$

is dense in H^1.

Proof Take $\varphi \in S$ satisfying $\varphi \equiv 1$ in a neighborhood of 0, φ has compact support. Let $\widehat{\psi} = \varphi$. Then $\psi \in S$ and $\int_{\mathbb{R}^n} \psi(x) \, dx = \varphi(0) = 1$. As usual, we let $\psi_\epsilon(x) = \epsilon^{-n}\psi(x/\epsilon)$. Thus $\widehat{\psi_\epsilon}(\xi) = \varphi(\epsilon\xi)$. For $h \in L^1$, we let $T_\epsilon(h) = h * \psi_\epsilon$. We claim that for $f \in H^1$, $T_\epsilon(I - T_\epsilon)f \to f$ in H^1 norm. To see this, note first that we must have $\widehat{f}(0) = 0$ if $f \in H^1$, otherwise $(-i\xi_j/|\xi|)\widehat{f}(\xi)$ could not be continuous at 0. If f has Fourier transform vanishing near 0 and near ∞, then $T_\epsilon(I - T_{1/\epsilon})f = f$ for ϵ sufficiently small. But functions with this property are trivially dense in the closed subspace of L^1 consisting of those f for which $\widehat{f}(0) = 0$. So we have shown that $T_\epsilon(I - T_{1/\epsilon}) \to I$ on a dense subspace of H^1. Since the operators $T_\epsilon(I - T_{1/\epsilon})$ are uniformly bounded in norm on L^1, it follows that $T_\epsilon(I - T_{1/\epsilon})f \to f$ in L^1 norm for all $f \in H^1$. But since these operators commute with the Riesz transforms, we get $T_\epsilon(I - T_{1/\epsilon})f \to f$ in H^1 norm for all $f \in H^1$. This proves Step α.

Claim β Given $f \in H^1$ with $\widehat{f}$ vanishing near $0, \infty$, and given $\eta > 0$, there exists a $g \in H_{00}^1$ so that $\|f - g\|_{H^1} < \eta$.

Proof of the Claim Given such an f, say $\widehat{f} = 0$ for $|\xi| < \delta$ or $|\xi| > 1/\delta$. We let $\varphi \in S$ be even, $\int_{\mathbb{R}^n} \varphi \, dx = 1$, φ supported in $\{|x| < 1\}$, $\varphi \equiv 1$ in a neighborhood of 0, $\varphi_\epsilon(x) = \epsilon^{-n}\varphi(x/\epsilon)$. And we let $\widehat{\psi_\epsilon} = \varphi_\epsilon$. Hence $\psi_\epsilon(x) = \psi_1(\epsilon x)$ so that $f_\epsilon(x) = \psi_\epsilon(x) \to f(x)$ in L^1 norm as $\epsilon \to 0$.

As soon as $\epsilon < \delta/2$, the Fourier transform of f_ϵ vanishes in a common neighborhood of 0 and ∞. Hence, when we look at

$$R_j f_\epsilon = \left(\frac{i\xi_j}{|\xi|}\widehat{f_\epsilon}\right)^{\vee},$$

we can redefine the multiplier off the common support of the f_ϵ so that it is C_c^∞. Let $\widehat{\tau}$ be the modified multiplier. So

$$\left(\frac{i\xi_j}{|\xi|}\widehat{f_\epsilon}(\xi)\right) = \widehat{\tau}(\xi)\widehat{f_\epsilon}(\xi)$$

whenever $\epsilon < \delta/2$ but $\widehat{\tau} \in C_c^\infty$. It follows that $\tau \in S$ and

$$R_j(f_\epsilon) = f_\epsilon * \tau \xrightarrow{L^1} f * \tau = R_j f \ , \ j = 1, 2, \ldots, n.$$

So $f_\epsilon \to f$ in H^1 as $\epsilon \to 0$. Of course each $f_\epsilon \in H_{00}^1$. This proves Claim β and Technical Point I.

For completeness, we include a proof of the fact that

$$S = \{f \in L^1 : \widehat{f} \text{ vanishes in some neighborhood of } 0\}$$

is dense in

$$\{f \in L^1 : \widehat{f}(0) = 0\}.$$

One can easily check that if suffices to assume that supp $f \subseteq \{|x| \leq M\}$, some $M > 0$. We may also assume that f is bounded and that $\int_{\mathbb{R}^n} f(x)\,dx = 0$. Let φ_ϵ be as in Claim β and $\psi_\epsilon = \epsilon^{-n}\psi(x/\epsilon)$, with $\widehat{\psi} = \varphi$. We show that $\|f * \psi_N\|_1 \to 0$ as $N \to \infty$ hence it follows that $f - f * \psi_N \in S$ and $f - f * \psi_N \to f$ in L^1 as desired.

Well,

$$|f * \psi_N(x)| = \left|N^{-n}\int_{\mathbb{R}^n} f(x-t)\psi(t/N)\,dt\right|$$

$$\leq N^{-n}\int_{|x-t|\leq M} |f(x-t)| \cdot |\psi(t/N) - \psi(x/N)|\,dt$$

since f has integral 0

$$\leq \|f\|_{L^\infty}N^{-n}\int_{|x-t|\leq M} |\psi(t/N) - \psi(x/N)|\,dt$$

$$\leq \|f\|_{L^\infty} \int_{|t| \leq M/N} |\psi(x/N) - \psi(x/N - t)| \, dt \, .$$

Thus

$$\|f * \psi_N\|_{L^1} \leq \|f\|_{L^\infty} \cdot N^n \int_{|t| \leq M/N} \omega_1(t) \, dt \, ,$$

where

$$\omega_1(t) = \int_{\mathbb{R}^n} |\psi(x - t) - \psi(x)| \, dx \, .$$

It is well known that $\omega_1(t) \to 0$ as $t \to 0$ so

$$N^n \int_{|t| \leq M/N} \omega_1(t) \, dt \to 0$$

as $N \to \infty$. This completes the proof. $\qquad\square$

PROBLEM For $p < 1$, are those elements of H^p which are smooth and are small at ∞ dense?

Technical Point II We shall need the notion of the Riesz transform of an L^∞ function. Let

$$K_j(x) = \frac{c_n x_j}{|x|^{n+1}} \quad \text{with } c_n = \frac{\Gamma((n+1)/2)}{\pi^{(n+1)/2}}$$

and, for $\epsilon > 0$, let

$$K_\epsilon^j(x) = \chi_{\{|x| > \epsilon\}}(x) \cdot K_j(x) \, .$$

Then the jth Riesz transform of $f \in L^\infty$ is defined to be

$$R_j f(x) = \lim_{\epsilon \to 0} \int_{\mathbb{R}^n} \left[K_\epsilon^j(x - y) - K_1^j(-y) \right] f(y) \, dy \, .$$

The easy estimate

$$|K_\epsilon^j(x - y) - K_1^j(-y)| \leq \frac{C|x|}{|y|^{n+1}}$$

for y large shows that the integral converges absolutely for every $\epsilon > 0$. The existence of the limit almost everywhere is clear once we know that

$$\lim_{\epsilon \to 0} \int_{\mathbb{R}^n} K_\epsilon^j(y) f(x-y) \, dy$$

exists almost everywhere for $f \in L^2$. This is easy by Plancherel. Then writing

$$f = f\chi_{\{|x|<1\}} + f\chi_{\{|x|\geq 1\}},$$

we notice that $\lim_{\epsilon \to 0}(R_\epsilon^j - R_1^j)(f\chi_{\{|x|\geq 1\}})$ is trivial and that $f\chi_{\{|x|<1\}}$ is an L^2 function, so we have defined R^j on L^∞.

Observe that, for $f \in L^\infty$ with compact support, $R^j f$ has been defined twice, and the two definitions differ by a constant. Finally, with reference to the theorem above and a limiting argument,

$$\|R^j f\|_* \leq C\|f\|_{L^\infty}.$$

Now we have the theorem of duality.

Theorem 5.7.8 (C. Fefferman) *If $f \in H_0^1$ and $\varphi \in BMO$, then the map*

$$f \mapsto \int_{\mathbb{R}^n} f(x)\varphi(x) \, dx$$

extends uniquely to a bounded linear functional on H^1. Moreover, every linear functional on H^1 arises in this way and the norm of the functional is comparable to $\|\varphi\|_{BMO}$.

Corollary 5.7.9 *The function $\varphi \in BMO$ if and only if*

$$\varphi = \varphi_0 + \sum_{j=1}^n R^j(\varphi_j)$$

for some $\varphi_j \in L^\infty$, $j = 0, 1, \ldots, n$.

Corollary 5.7.10 *The function $\varphi \in BMO$ if and only if*

$$\int_{\mathbb{R}^n} \frac{|\varphi(x)| \, dx}{(1+|x|)^{n+1}} < \infty$$

and

$$\iint_{T(x_0,h)} |\nabla\varphi(x,y)|^2 y \, dy \leq Ch^n,$$

where $x_0 \in \mathbb{R}^n$, $h > 0$, and

$$\varphi(x, y) = PI\varphi \quad, \quad T(x, h) = \{(x, y) : |x - x_0| < h, 0 < y < h\}.$$

Let us number these statements as follows:

(1) The mapping $f \mapsto \int f\varphi$ for $f \in H_0^1$ extends to a continuous linear functional on H^1.

(2) We have $\varphi = \varphi_0 + \sum_j R^j \varphi_j$, $\varphi_j \in L^\infty$.

(3) $\varphi \in BMO$.

(4) The condition in the second corollary above.

We will prove **(1)** $\Rightarrow$ **(2)** $\Rightarrow$ **(3)** $\Rightarrow$ **(4)** $\Rightarrow$ **(1)**.

Proof The implication **(1)** $\Rightarrow$ **(2)** is generalized nonsense. Let $B = L^1 \oplus L^1 \oplus \cdots \oplus L^1$, with there being $(n + 1)$ factors, and set

$$\|f\|_B = \sum_j \|f_j\|_{L^1}$$

for $f = (f_0, f_1, \ldots, f_n) \in B$. Let $S = \{f \in B : f_j = R^j f_0, j = 1, 2, \ldots, n\}$. It is easy to see that S is a closed, linear subspace of B and $S \approx H^1$. If $\mathcal{L}$ is a continuous linear functional on H^1, then $\mathcal{L}$ is a continuous linear functional on S. So $\mathcal{L}$ extends to B by the Hahn–Banach theorem. Hence there exist $\varphi_0, \varphi_1, \ldots, \varphi_n \in L^\infty$ such that

$$L(f_0, f_1, \ldots, f_n) = \sum_{j=0}^{n} \int f_j \varphi_j \, dx$$

$$= \int f_0 \varphi_0 \, dx + \sum_{j=1}^{n} \int f_j \varphi_j \, dx$$

$$= \int f_0 \varphi_0 \, dx + \sum_{j=1}^{n} \int R^j(f_j) \varphi_j \, dx$$

$$= \int f_0 \varphi_0 \, dx + \sum_{j=1}^{n} \int f_0 \cdot R^j(-\varphi_j) \, dx$$

$$= \int f_0 \left[\varphi_0 + \sum_j R^j(-\varphi_j) \right] dx.$$

So if we assume **(1)**, then $\varphi \in BMO$ implies that there exists $\varphi_0, \varphi_1, \ldots \varphi_n$ such that

$$\int f\varphi \, dx = \int f \left[\varphi_0 + \sum_j R^j(\varphi_j) \right] dx \quad \text{for all } f \in H_0^1 .$$

Hence $\varphi = \varphi_0 + \sum_j R^j \varphi_j$.

$(2) \Rightarrow (3)$ has been proved.

For $(3) \Rightarrow (4)$, we know that if $\varphi \in BMO$ and Q is the unit cube centered at 0, then

$$\int \frac{|\varphi - \varphi_Q|}{(1+|x|)^{n+1}} \, dx \leq C \|\varphi\|_* .$$

By a change of variables, if $\delta > 0$ and Q_δ is the cube centered at 0 of side δ, then

$$\int \frac{|\varphi - \varphi_{Q_\delta}|}{(1+|x|)^{n+1}} \, dx \leq \frac{C}{\delta} \|\varphi\|_* .$$

Let us now fix $\varphi \in BMO$, $h > 0$, and suppose that $x_0 = 0$. Let Q_{4h} be the cube of side $4h$ and centered at 0. Let $\chi = \chi_{Q_{4h}}$ and $\widetilde{\chi} = 1 - \chi$. Thus we write

$$\varphi(x) = \varphi_{Q_{4h}} + (\varphi - \varphi_{Q_{4h}})\chi + (\varphi - \varphi_{Q_{4h}})\widetilde{\chi} = \varphi_1 + \varphi_2 + \varphi_3 .$$

Hence $\varphi(x,t) = \varphi_1(x,t) + \varphi_2(x,t) + \varphi_3(x,t)$. Now $\nabla \varphi_1 = 0$ on $T(x,h)$ so that there is nothing to estimate for φ_1. We have

$$\int_{T(0,h)} y|\nabla\varphi_2|^2 \, dxdy \leq \int_{\mathbb{R}^{n+1}_+} y|\nabla\varphi_2|^2 \, dxdy = \int_{\mathbb{R}^n} S(u)^2(x) \, dx$$

by an identity we have used when we studied the area integral S. But this is

$$\leq C\|\varphi_2\|_2^2$$

$$\leq C \int_{Q_{4h}} |\varphi - \varphi_{4h}|^2 \, dx$$

$$\leq C|Q_{4h}| \|\varphi\|_{BMO}^2$$

$$\leq C' h^n \|\varphi\}_{BMO}^2 .$$

Finally,

$$|\nabla\varphi_3(x,y)| \leq \int |\nabla P_y(x-t)| |\varphi_3(t)| \, dt$$

$$\leq A \int_{^cQ_{4h}} \left[\frac{1}{y+|x-t|} \right]^{n+1} \cdot |\varphi(t) - \varphi_{Q_{4h}}| \, dt$$

Now if $(x, y) \in T(0, h)$, then

$$\left(\frac{1}{y + |x - t|}\right)^{n-1} \leq \frac{A}{h^{n+1} + |t|^{n+1}}$$

when $t \in {}^c Q_{4h}$. Thus

$$|\nabla \varphi_3(x, y)| \leq \frac{A}{h} \|\varphi\|_*$$

by the remark at the beginning of the proof of **(3)** $\Rightarrow$ **(4)**.
      Now

$$\int_{T(0,h)} |\nabla \varphi_3(x, y)|^2 y \, dx dy \leq \int_{\substack{|x| \leq h \\ |y| \leq h}} \frac{A^2}{h^2} \|\varphi\|_*^2 y \, dx dy$$

$$\leq A_0 h^n \|\varphi\|_*^2 .$$

Thus a combination of the estimates on φ_1, φ_2, φ_3 gives the desired result.   □

   There is a famous problem related to this discussion. Namely, one can ask to characterize those Borel regular measures $d\mu$ on $\mathbb{R}_+^{n+1}$ such that if $u(x, y) = PI(f)$, $f \in L^p(\mathbb{R}^n)$, $1 < p < \infty$, then

$$\iint_{\mathbb{R}_+^{n+1}} |u(x, y)|^p \, d\mu(x, y) \leq C \int_{\mathbb{R}^n} |f(x)|^p \, dx .$$

It turns out that a measure μ satisfies such an inequality for all $f \in L^p$ if and only if

$$\iint_{\Gamma(x_0)} d\mu \leq C h^n \text{ for all } 0 < h < \infty .$$

Outline of the Proof For the necessity, use $u(x, y) = P_{y+h}(x) = PI(P_h(x))$. For the sufficiency, one proves that if u^* is the nontangential maximal function of $u(x, y)$ then, for any p,

$$\int |u(x, y)|^p \, d\mu \leq C_p \int u^*(x)^p \, dx .$$

This certainly does it. Carleson's original proof was very ingenious and took place on the disc. Hörmander did it in the n-dimensional case. The whole thing turns out to be a consequence of the Hardy–Littlewood maximal theorem. Let us see how the proof of necessity works for $p = 1$, $n = 1$, anyway: We claim that

Fig. 5.14 A triangle erected over each open interval

$$\mu\{(x, y) : |u(x, y)| > \alpha\} \le Cm\{x : u^*(x) > \alpha\}$$

if μ is a Carleson measure.

First note that $\{u^*(x) > \alpha\}$ is open. On the line, open sets are disjoint unions of open intervals. We erect a triangle of aperture 1 over each of these open intervals. See Fig. 5.14.

Clearly $|u| \le \alpha$ outside the union of the shaded triangles. Thus

$$\mu\{(x, y) : |u(x, y)| > \alpha\} \le \sum_{\text{cones}} \mu(\text{cone}) \le \sum_{\text{boxes}} \mu(\text{box}),$$

where we set each cone (or triangle) in a box of roughly the same base and height and not more than 4 times the area. The last expression is $\le C \sum m|\text{bases}|$, where m is one-dimensional Lebesgue measure. But this is

$$\le 4Cm\{x : u^*(x) > \alpha\}$$

as desired.   □

Now, for $n > 1$, a modification is needed. Namely, we use the Whitney decomposition of an open set.

The Whitney Decomposition There exist two constants C_1, C_2 such that if Ω is any open subset of $\mathbb{R}^n$, then there is a disjoint collection of closed cubes in Ω (by "disjoint" we mean disjoint interiors) such that $\cup_k Q_k = \Omega$ and, for all k,

$$C_1 \cdot \text{diam } Q_k \le \text{dist}(Q_k, \partial\Omega) \le C_2 \cdot \text{diam } Q_k.$$

Once we have this lemma, the argument for $\mathbb{R}^n$ goes through with only obvious modifications.

Proof of the Whitney Decomposition Let $\mathcal{M}_0$ be the decomposition of $\mathbb{R}^n$ into disjoint cubes with vertices in $\mathbb{Z}^n$ and let $\mathcal{M}_k = 2^k \mathcal{M}_0$ in the obvious sense, $-\infty < k < \infty$. We deem a cube "admissible" if its diameter is proportional to its distance from $\partial\Omega$ in the sense we have indicated. Every point of Ω is contained in the interior of an admissible cube. So the acceptable cubes cover Ω. But we wish to have a disjoint subcollection. So we let $C_1 = 1$, $C_2 = 4$, and observe that if two cubes Q_1, Q_2 from $\mathcal{M}_{k_1}$ and $\mathcal{M}_{k_2}$ have nontrivial intersection, then one of them is contained in the other. Thus if, for each $x \in \Omega$, we select the "largest" admissible cube containing it, it is easy to see that these "largest" cubes provide a cover of Ω with disjoint interiors. For further details, see [STE1].

Having disposed of Carleson measures, we return to **(4)** $\Rightarrow$ **(1)**. Our hypothesis about φ says that the measure $d\mu = y|\nabla\varphi(x, y)|^2 \, dx \, dy$ is a Carleson measure. Now suppose that $F \in H^1$ so that

$$F = (u_0(x, y), u_1(x, y), \ldots, u_n(x, y))$$

with

$$u_0 = PI(f) \ , \quad u_j = PI(R^j f) \ , \quad \text{some } f \in L^1 \ .$$

We wish to see that $\int_{\mathbb{R}^{n+1}_+} |F| \, d\mu \le C\|F\|_{H^1}$. This is a restatement of **(1)**.

We recall the identity

$$\int_{\mathbb{R}^n} |f(x)|^2 \, dx = 2 \int_{\mathbb{R}^{n+1}_+} |\nabla u(x, y)|^2 y \, dx \, dy$$

for f which are appropriately bounded at ∞. The polarized form of this identity is

$$\int_{\mathbb{R}^n} f(x)\varphi(x) \, dx = 2 \int_{\mathbb{R}^{n+1}_+} \nabla u_0(x, y) \cdot \nabla\varphi(x, y) y \, dx \, dy \ .$$

Thus

$$\left| \int_{\mathbb{R}^n} f(x)\varphi(x) \, dx \right| \le 2 \int_{\mathbb{R}^{n+1}_+} |\nabla u_0(x, y)||\nabla\varphi(x, y)| y \, dx \, dy \ .$$

But

$$|\nabla F| \ge |\nabla u_0|$$

hence this is

$$\le 2 \int_{\mathbb{R}^{n+1}_+} |\nabla F||\nabla\varphi| y \, dx \, dy \ .$$

Now recall that

$$\Delta(|F|^p) \ge C_p |\nabla F|^2 |F|^{p-2} \ , \quad p > (n-1)/n \ .$$

For $p = 1$, this says that $\Delta(|F|) \ge C_1 |\nabla F|^2/|F|$. One must be careful about the set where F vanishes, but we momentarily ignore this problem. So

$$|\nabla F|^2 \le C|F|\Delta(|F|) \ .$$

We also need the identity

$$\int_{\mathbb{R}^{n+1}_+} y(\triangle|F|)\, dxdy + \int_{\mathbb{R}^n} |F(x,0)|\, dx \le C\|f\|_{H^1}.$$

Putting all this together, we notice that

$$\left| \int_{\mathbb{R}^n} f(x)\varphi(x)\, dx \right| \quad \le \quad C \int_{\mathbb{R}^{n+1}_+} (|F|\,|\triangle|F||)^{1/2} |\nabla\varphi| y\, dxdy$$

$$\overset{\text{(Schwarz)}}{\le} \quad C \int_{\mathbb{R}^{n+1}_+} |\triangle|F||y\, dxdy^{1/2} \int_{\mathbb{R}^{n+1}_+} |F|y|\nabla\varphi|^2\, dxdy^{1/2}$$

$$\le \quad C\|f\|_{H^1}^{1/2} \cdot \|f\|_{L^1}^{1/2}$$

$$\le \quad C'\|f\|_{H^1}.$$

This is the desired inequality. One can get around the zeros of F by instead doing estimates on $F + \epsilon$ (a fixed Poisson integral) and then letting $\epsilon \to 0$.

Finally, we show that $\varphi \in BMO$ and $\int \varphi h = 0$ for all $h \in H^1$ implies that $\varphi \equiv 0$. One can check that, for $\delta > 0$ fixed, $(\partial P_\delta / \partial x_j)(x)$ is an H^1 function. Thus

$$\int_{\mathbb{R}^n} \varphi(t) \frac{\partial}{\partial x_j} P_\delta(x - t)\, dt \equiv 0 \quad \text{for all } j = 1, 2, \dots, n.$$

So if $u = PI\varphi$, this says that u is a function of y only. But $(\partial P_y / \partial y)(x)$ is also in H^1 so the same argument shows that u is constant. Hence φ is constant. So it must be 0. [Note that one can use derivatives of P rather than P itself since the latter is not small enough at ∞.]

The following characterization of $H^1(\mathbb{R}^1)$ is due to Coifman and Herz (see [COI] and [HER]. A function $a : \mathbb{R}^1 \to \mathbb{R}^1$ is said to be an *atom* for the space H^1 if a is supported on an interval I, $\int_I a(x)\, dx = 0$, and $|a(x)| \le 1/|I|$. One has

Theorem 5.7.11 *A function f lies in $H^1(\mathbb{R}^1)$ if and only if*

$$f = \sum_k c_k a_k(x)$$

with the a_k atoms and $\sum_k |c_k| < \infty$. In fact there exist K_1, K_2 positive such that

$$K_1 \sum_j |c_j| \le \|f\|_{H^1} \le K_2 \sum_j |c_j|.$$

This result makes it very easy to pair BMO functions with H^1 functions. For if $\varphi \in BMO$, then

$$\left| \int f\varphi \, dx \right| \le \left| \sum_k c_k \int a_k(x)\varphi(x) \, dx \right|$$

$$\le \sum_k |c_k| \frac{1}{m(I_k)} \int_{I_k} |\varphi(x) - \varphi_{I_k}| \, dx$$

$$\le \left(\sum_k |c_k| \right) \|\varphi\|_* .$$

An analogous result is true for $n = 1$, $p < 1$. We remark that this theorem is related to Riesz's theorem on the rising slope.

OPEN PROBLEM Extend the notion of atoms to $\mathbb{R}^n$, $n > 1$.

Problem Return to non-isotropic dilations. One can define BMO in this setting. Is BMO now the dual of some H^1 space? Can one isolate the right Riesz transforms? Can one show that

$$f \in BMO \quad \text{if and only if} \quad f = f_0 + \sum R^j f_j , \quad f_j \in L^\infty ?$$

More specifically, can one do this theory on the Heisenberg group? Here the Heisenberg group (simplest case) can be thought of as

$$\left\{ \begin{pmatrix} 1 & x_1 & x_3 \\ 0 & 1 & x_2 \\ 0 & 0 & 1 \end{pmatrix} \right\}$$

under matrix multiplication. (See [FOS] and [KRA1] for more on the Heisenberg group.) And the appropriate dilations, which are automorphisms of the group, are

$$\alpha_\delta : \begin{pmatrix} 1 & x_1 & x_3 \\ 0 & 1 & x_2 \\ 0 & 0 & 1 \end{pmatrix} \longmapsto \begin{pmatrix} 1 & \delta x_1 & \delta^2 x_3 \\ 0 & 1 & \delta x_2 \\ 0 & 0 & 1 \end{pmatrix} .$$

Chapter 6
Developments Since 1974

Nearly fifty years have passed since the conclusion of Stein's 1973–1974 course. Since those years marked the very beginning of the modern theory of real variable Hardy spaces, it is natural to expect that there have been many new developments. In the present chapter we shall briefly describe just some of the landmarks.

6.1 The Atomic Theory

One of the most decisive new ideas that has come about since the conclusion of Stein's class is the atomic decomposition for Hardy spaces. Based on a remark of C. Fefferman, the first paper in the subject was [COI]. This was quickly followed by [LAT]. The idea of atoms was brought into full bloom by the more substantial paper [COW]. We now give some definitions and state a few definitive results.

A function $a : \mathbb{R}^N \to \mathbb{C}$ is called a *1-atom* if

- a is supported in a ball $B(P, r)$.
- $|a(x)| \leq \frac{1}{|B(P,r)|}$.
- $\int a(x)\, dx = 0$.

The basic theorem about these atoms is as follows:

Theorem 6.1.1 *A function $f : \mathbb{R}^N \to \mathbb{C}$ is in H^1 if and only if*

$$f = \sum_j \alpha_j a_j,$$

where each a_j is a 1-atom and $\{\alpha_j\} \in \ell^1$ and

$$\|f\|_{H^p} \approx \|\{\alpha_j\}\|_{\ell^1}.$$

© The Author(s), under exclusive license to Springer Nature Switzerland AG 2023   227
S. G. Krantz, *The E. M. Stein Lectures on Hardy Spaces*, Lecture Notes in
Mathematics 2326, https://doi.org/10.1007/978-3-031-21952-8_6

It is easy to see that this theorem is easily generalized to spaces of homogeneous type (see [COW]).

Matters are a bit more subtle for H^p with $0 < p < 1$. For such p we define $a : \mathbb{R}^N \to \mathbb{C}$ to be a p-atom if and only if

- a is supported in a ball $B(P, r)$.
- $|a(x)| \leq |B(P, r)|^{-1/p}$.
- $\int a(x) x^\alpha \, dx = 0$ for any multi-index α satisfying $|\alpha| \leq [N \cdot (1/p - 1)]$.

Here [ ] denotes the greatest integer function.

The main theorem about H^p for $0 < p < 1$ is this:

Theorem 6.1.2 *Let* $0 < p < 1$. *A function* $f : \mathbb{R}^N \to \mathbb{C}$ *is in* H^p *if and only if*

$$f = \sum_j \alpha_j a_j,$$

where each a_j *is a p-atom and* $\{\alpha_j\} \in \ell^p$. *The series is assumed to converge in the distribution sense. Furthermore,*

$$\|f\|_{H^p} \approx \|\{\alpha_j\}\|_{\ell^p}.$$

It is easy to see that the definition of H^p for p less than 1 but very near 1 makes sense on a space of homogeneous type—just because the mean value condition only involves orthogonality to constants. But the definition of H^p for smaller p makes no sense on a general space of homogeneous type (although see [KRA2, KRA3] for some ideas about how to address this issue).

The atomic theory of H^p spaces is today very well developed. It should be noted, for instance, that singular integral operators *do not* map atoms to atoms. For this reason the theory of molecules was developed by M. Taibleson and others (see [TAW, CRTW]).

6.2   The Local Theory of Hardy Spaces

In his thesis (see, for instance, [GOL]), David Goldberg developed a local theory of Hardy spaces. This theory was, in turn, based in part on the theory of atoms. The local Hardy spaces are denoted by $h^p(\mathbb{R}^N)$. Among the many advantages of this local theory are (**i**) the Schwartz space is a dense subspace of h^p, (**ii**) pseudodifferential operators of order 0 are bounded on h^p, (**iii**) h^p is closed under multiplication by smooth cutoff functions, and (**iv**) h^p makes sense on manifolds. Let us now provide a few of the details (without proofs).

Fix a dimension $N \geq 1$. Define

$$S = \{(x_1, x_2, \ldots, x_N, y) \in \mathbb{R}^{N+1} : 0 < y < 1\}.$$

We shall refer to S as the *strip* in $\mathbb{R}^{N+1}$. It is easy to see that the Poisson kernel P_y for S exists (see [KRA4]).

Now, if $p > (N-1)/N$, then set

$$h^p \equiv \{(u_0, u_1, u_2, \ldots, u_N) :$$

$$\sup_{0 < y < 1} \int_{\mathbb{R}^N} \left(|u_0(x, y)|^2 + |u_1(x, y)|^2 + \cdots + |u_N(x, y)|^2 \right)^{p/2} dx < \infty \},$$

where each u_j is harmonic on S and the $\{u_j\}$ satisfy the generalized Cauchy–Riemann equations

$$\sum_{j=0}^{N} \frac{\partial u_j}{\partial x_j} = 0 \quad , \quad \frac{\partial u_j}{\partial x_k} = \frac{\partial u_k}{\partial x_j} \quad \text{for } j \neq k.$$

We impose the additional condition $u_0(x, y) = u_0(x, 1-y)$. Compare this definition with the one in [STW2]. The main thing to notice is that we replace the upper half plane with the strip S. As in the classical theory of H^p spaces, $u_1, u_2, \ldots, u_N$ can be recovered from u_0 by way of certain Riesz transforms r_j (see below).

Now let u be a harmonic function on S. For $x \in \mathbb{R}^N$ define $\Gamma(x) = \{(x', y) : |x - x'| < y, 0 < y < 1/2\}$. This is a standard cone, truncated in the y variable. Set

$$u^*(x) = \sup_{(x', y) \in \Gamma(x)} |u(x', y)|.$$

Let $\mathcal{S}$ denote the Schwartz space and let $\|\varphi\|_{\alpha\beta} = \|x^\alpha D^\beta \varphi\|_\infty$. We will assume that our harmonic function u on the strip S is the Poisson integral of a function f on the boundary of S. Finally, if g is any function on $\mathbb{R}^N$, then let $g_t(x) = t^{-N} g(x/t)$. Then we have

Theorem 6.2.1 *If $p > (N-1)/N$, then $u \in h^p$ if and only if $u^* \in L^p$. For each $0 < p < \infty$ there exists an M so that if*

$$B = \{\varphi \in \mathcal{S} : \|\varphi\|_{\alpha\beta} \leq 1 \text{ for } |\alpha|, |\beta| \leq M\},$$

and if $\psi \in \mathcal{S}$ with $\int \psi \neq 0$, then the L^p norms of the following functions are equivalent:

- u^*
- $\sup_{(x',t) \in \Gamma(x)} |\psi_t * f(x')|$
- $\sup_{0 < t < 1} |\psi_t * f(x)|$
- $\sup_{0 < t < 1} \sup_{\varphi \in B} |\varphi_t * f(x)|$

We have alluded above to the Riesz transforms in this new context. Let us say what they are. Let $\mu \in C_c^\infty(\mathbb{R}^N)$ be identically equal to 1 in a neighborhood of the origin. Define

$$(r_j f)\widehat{} = (1 - \mu) \cdot i \cdot \frac{\xi_j}{|\xi|} \cdot \widehat{f}.$$

Then a distribution f belongs to h^1 if and only if $f \in L^1$ and $r_j f \in L^1$ for $j = 1, 2, \ldots, N$.

Now let us define the local space of functions of bounded mean oscillation.

Definition 6.2.2 Let b be a locally integrable function on $\mathbb{R}^N$. We say that b is *locally of bounded mean oscillation* and write $b \in$ BMO if

$$\sup_{|Q|<1} \frac{1}{|Q|} \int_Q \left| b - \frac{1}{|Q|} \int_Q b \right| dx < \infty$$

and

$$\sup_{|Q|>1} \frac{1}{|Q|} \int_Q |b| \, dx < \infty.$$

Here Q is a cube in $\mathbb{R}^N$ with sides parallel to the axes and $|Q|$ is the volume or measure of Q.

Then there is a Fefferman–Stein decomposition for BMO functions. Namely, $b \in$ BMO if and only if there exist $b_0, b_1, \ldots, b_N \in L^\infty$ such that

$$b = b_0 + r_1 b_1 + r_2 b_2 + \cdots + r_N b_N.$$

It is also the case that the dual of h^1 is BMO.

Now let us say a few words about the atomic decomposition for h^p functions. [This atomic decomposition was important for the work of Chang/Krantz/Stein that is described in the next section.] Fix p. We define a p-atom to be a function a such that

- a is supported in a ball $B(P, r)$.
- $|a(x)| \leq 1/|B(P, r)|$.
- $\int a(x) x^\alpha \, dx = 0$ for $|\alpha| \leq [N(p^{-1} - 1)]$ and $|B(P, r)| \leq 1$.

Notice that there is no moment condition in case the smallest ball containing the support of a has volume ≥ 1. We define $\|a\|_{\text{atom}}^p \equiv \|a\|_\infty^p \cdot |B(P, r)|$. Note that $\|a\|_{h^p} \leq C \|a\|_{\text{atom}}$. Then we have

Theorem 6.2.3 *If $p \leq 1$ and $f \in h^p$, then $f = \sum_j a_j$ with $\sum_j \|a_j\|_{\text{atom}}^p \leq C\|f\|_{h^p}^p$. In fact*

$$f = \lim_{k \to \infty} \sum_{j=1}^k a_j$$

in the h^p norm.

The following result tells us that the local Hardy spaces make sense on a manifold:

Proposition 6.2.4 *If $\lambda : \mathbb{R}^N \to \mathbb{R}^N$ is a C^∞ diffeomorphism such that $\lambda(x) = x$ for $|x| > 1$, then*

$$\| f \circ \lambda \|_{h^p} \leq C \| f \|_{h^p} .$$

We conclude this discussion with a few words about the dual space of h^p when $p < 1$. For $0 < \alpha < \infty$, let Λ_α be the usual Lipschitz space of bounded functions with bounded derivatives (see [KRA5] for a thorough discussion of these spaces). Then the dual of h^p for $0 < p < 1$ is $\Lambda_{N(p^{-1}-1)}$.

6.3   The Work of Chang/Krantz/Stein on Hardy Spaces for Elliptic Boundary Value Problems

The original theory of real variable Hardy spaces, as developed by Stein and Weiss in [STW2] and later by Fefferman and Stein in [FES], was adapted to the study of the Laplacian on all of Euclidean space. It is natural to ask whether there is a theory of real variable Hardy spaces for elliptic boundary value problems on a smoothly bounded domain in Euclidean space. In fact there is such a theory, and it was developed by Chang, Krantz, and Stein in [CKS1, CKS2].

Fix a smoothly bounded domain $\Omega \subseteq \mathbb{R}^N$. We consider two notions of $H^p(\Omega)$, one of which is in a sense the largest possible Hardy space, and the other of which is the smallest. The largest, $H_r^p(\Omega)$, arises by restricting to Ω arbitrary elements of $H^p(\mathbb{R}^N)$. The other arises by restricting to Ω the elements of $H^p(\mathbb{R}^N)$ which are zero outside of $\overline{\Omega}$. We call this second space $H_z^p(\Omega)$. More precisely, we will use the localized Hardy spaces of Goldberg which were described in the last section. In particular, the localized theory allows for multiplication by smooth cutoff functions. Thus the spaces we use are $h_r^p(\Omega)$ and $h_z^p(\Omega)$.

The Dirichlet problem that is considered is in finding the $u = G(f)$ which solves $\Delta u = f$ in Ω with $u\big|_{\partial\Omega} = 0$. The Neumann problem is to find the $u = \widetilde{G}(f)$ which solves $\Delta u = f$ in Ω with $(\partial u/\partial \nu)\big|_{\partial\Omega} = 0$ (where $\partial/\partial \nu$ is the outward normal derivative).

One of the first main results is that the mappings

$$f \longmapsto \frac{\partial^2}{\partial x_j \partial x_k} G(f)$$

and

$$f \longmapsto \frac{\partial^2}{\partial x_j \partial x_k} \widetilde{G}(f)$$

are both bounded from $h_z^p(\Omega)$ to $h_r^p(\Omega)$ for all $0 < p \leq 1$.

For the Dirichlet problem, the mapping

$$f \longmapsto \frac{\partial^2}{\partial x_j \partial x_k} G(f)$$

extends to a bounded mapping from $h_r^p(\Omega)$ to $h_r^p(\Omega)$ when $N/(N+1) < p \leq 1$. Simple examples show that we *do not* have bounded extensions from $h_r^p(\Omega)$ to $h_r^p(\Omega)$ for the Neumann problem for any $p \leq 1$. In addition, boundedness fails for the Dirichlet problem whenever $p \leq N/(N+1)$.

6.4 Multi-Parameter Harmonic Analysis

Harmonic analysis for operators associated with multi-parameter families of dila-tions is still very much in its infancy—little is known beyond that which follows from the product theory. Let MZ be the maximal operator and TZ be any of certain singular integral operators which are naturally associated with the two-parameter family of Zygmund dilations. It has been proved that MZ and TZ are bounded on $L^p(w)$, $1 < p < \infty$, for all weights w in the correct version of Muckenhoupt's A^p class. Boundedness of these operators on unweighted L^p was already known, reducing in one way or another to the product theory; these results constitute an important advance because such a reduction is impossible in the weighted case. See [FEP] for the details.

In a related vein, work of R. Fefferman [RFEF] endeavors to find to what extent the duality between the Hardy space H^1 of the unit disc D in the complex plane and the space BMO (of functions of bounded mean oscillation) extends to the polydisc. There are two parts to the reasoning. The first is a counterexample, which indicates that the space defined by a natural generalization of the one-variable BMO definition fails to act on H^1 of the polydisc. The example that is constructed is based on the famous Carleson counterexample concerning the boundedness of (Carleson) measures on H^p for the bidisc (see [CAR]). The second part contains a necessary and sufficient condition for a function to be in the dual space H^1 of the bidisc. See also [CHF] for related investigations. The paper [RFES] considers singular integrals in the multi-parameter setting.

The paper [MER] explores Hardy space theory for $p < 1$ in the multi-parameter setting.

6.5 The $T1$ Theorem of David/Journé

Let $T : L^2(\mathbb{R}^N; dx) \to L^2(\mathbb{R}^N; dx)$ be linear and continuous. We say that T is a Calderón–Zygmund operator if the following properties are satisfied: there exists a function $K(x, y)$, locally integrable on the open set Ω of $\mathbb{R}^N \times \mathbb{R}^N$ consisting of those pairs (x, y) for which $y \neq x$, and such that **(1)** $Tf(x) = \int_{\mathbb{R}^N} K(x, y) f(y) \, dy$ whenever x does not belong to the compact support of $f \in C_c^\infty(\mathbb{R}^N)$, **(2)** $\int_{|y-x| \geq 2|x'-x|} |K(x', y) - K(x, y)| \, dy \leq C$ and **(3)** $\int_{|y-x| \geq 2|y'-y|} |K(x, y') - K(x, y)| \, dx \leq C$.

For such operators, the real variable methods introduced by A. P. Calderón and A. Zygmund apply and provide L^p-boundedness for $1 < p < +\infty$. Moreover T is bounded from $L^\infty(\mathbb{R}^N)$ into $BMO(\mathbb{R}^N)$, the space introduced by John and Nirenberg. In the applications, **(2)** and **(3)** are easily checked. On the contrary, the problem of finding an efficient criterion for the L^2-boundedness of a weakly defined operator $T : C_c^\infty \to (C_c^\infty)^*$ whose kernel satisfies **(2)** and **(3)** is still open.

In the paper [DAJ], the authors replace **(2)** and **(3)** by the stronger assumptions: **(2')** there exists $\delta > 0$ such that $|K(x', y) - K(x, y)| \leq C|x' - x|^\delta |x - y|^{-N-\delta}$ whenever $|x' - x| \leq (1/2)|x - y|$; **(3')** a similar estimate holds for $|K(x, y') - K(x, y)|$ when $|y' - y| \leq (1/2)|x - y|$.

Then the following remarkable theorem holds: For a weakly defined operator $T : C_c^\infty \to (C_c^\infty)^*$ whose kernel satisfies **(2')** and **(3')**, the L^2-boundedness of T is equivalent to the requirements that $T(1)$ should belong to BMO, that $T^*(1)$ should also belong to BMO, and that T should enjoy the weak boundedness property. The meaning of $T(1)$ is as follows. If $\varphi \in C_c^\infty(\mathbb{R}^N)$ and $\varphi(0) = 1$, φ_ϵ is defined by $\varphi_\epsilon(x) = \varphi(\epsilon x)$. Then there exist constants c_ϵ (which may blow up as ϵ tends to 0) such that $T(\varphi_\epsilon) = c_\epsilon + R_\epsilon$, where R_ϵ converges in the distributional sense to an object named $T(1)$ and defined modulo constant functions. The operator T^* is a weak adjoint of T and $T^*(1)$ is, therefore, also a distribution modulo a constant function. The weak boundedness property means the existence of an integer $m \geq 1$ and a constant C such that for each cube Q and each pair φ_1, φ_2 of two testing functions supported by Q we have $|\langle T(d_Q), \varphi_2 \rangle| \leq C p_{m,Q}(\varphi_1) p_{m,Q}(\varphi_2)|Q|$, where $|Q|$ is the measure of Q and where (d_Q being the diameter of Q) $p_{m,Q}(\varphi) = \sup_{|\alpha| \leq m} \sup_Q |\partial^\alpha \varphi| d_Q^{|\alpha|}$.

The references [JOUR1, JOUR2] explore applications of these ideas to multi-parameter harmonic analysis.

6.6 Contributions of Tom Wolff

Thomas Wolff was a remarkable mathematician who made many important contributions to analysis. His first was a new proof of Carleson's corona theorem based on harmonic analysis ideas. Wolff never published his proof, but it can be found in [KOO] or [KRA1].

The papers   [WOL1, WOL2] contribute to our understanding of the role of the Kakeya set in harmonic analysis. In particular, his ideas have applications to Bochner–Riesz means and to the wave equation.

Perhaps one of Wolff's most dramatic results appears in [WOL3]. Here he shows that the Cauchy–Riemann equations approach of Stein and Weiss (see [STW2]) cannot be used to characterize real variable $H^p(\mathbb{R}^N)$ when $p < (N - 1)/N$.

It is tragic that Wolff died an early death in 2000.

6.7   Wavelets

Certainly one of the big developments in harmonic analysis in the past forty years has been the theory of wavelets. Largely due to Yves Meyer, this is a new type of Fourier analysis that is *not* based on trigonometric functions. It is in fact a much more flexible and powerful theory that allows localization in both the space and the phase variables. A good introduction to wavelet theory appears in [LWW]. A more comprehensive introduction is in [HERG].

While wavelets are not literally a part of the real variable theory of Hardy spaces, they are closely related to the atomic decomposition. This assertion is especially clear in the theory of the φ-transform (a set of ideas analogous to wavelets), as represented in [FRJ].

We present here a brief description of what wavelet theory is about.

6.7.1   *Localization in the Time and Space Variables*

The premise of the new versions of Fourier analysis that are being developed today is that sines and cosines are not an optimal model for some of the phenomena that we want to study. As an example, suppose that we are developing software to detect certain erratic heartbeats by analysis of an electrocardiogram. [Note that the discussion that we present here is philosophically correct but is over-simplified to facilitate the exposition.] The scheme is to have the software break down the patient's electrocardiogram into component waves. If a wave that is known to be a telltale signal of heart disease is detected, then the software notifies the user.

A good plan, and there is indeed software of this nature in use across America. But let us imagine a typical electrocardiogram and further imagine an aberrant heartbeat that we wish to detect.

What we want the software to do is to break up the electrocardiogram wave into fundamental components, and then to see whether one of those components is the aberrant wave. Of what utility is Fourier theory in such an analysis? Fourier theory would allow us to break the electrocardiogram wave into sines and cosines, then break the aberrant wave into sines and cosines, and then attempt to match up coefficients. Such a scheme may be dreadfully inefficient, because sines and cosines *have nothing to do* with the waves we are endeavoring to analyze.

The Fourier analysis of sines and cosines arose historically because sines and cosines are eigenfunctions for the wave equation. Their place in mathematics became even more firmly secured because they are orthonormal in L^2. They also commute with translations in natural and useful ways. The standard trigonometric relations between the sine and cosine functions give rise to elegant and useful formulas—such as the formulas for the Dirichlet kernel and the Fejér kernel and the Poisson kernel. Sines and cosines have played an inevitable and fundamental historical role in the development of harmonic analysis.

. In the same vein, translation-invariant operators have played an important role in our understanding of how to analyze partial differential equations (see [KRA6]), and as a step toward the development of the more natural theory of pseudodifferential operators. Today we find ourselves studying translation *non-invariant* operators—such as those that arise in the analysis on the boundary of a (smoothly bounded) domain in $\mathbb{R}^N$. The $T(1)$ theorem of David-Journé gives the most natural and comprehensive method of analyzing integral operators, and their boundedness on a great variety of spaces.

The next, and current, step in the development of Fourier analysis is to replace the classical sine and cosine building blocks with more flexible units—indeed, with units that can be tailored to the situation at hand. Such units should, ideally, be localizable; in this way they can more readily be tailored to any particular application. This, roughly speaking, is what wavelet theory is all about.

In a brief treatment of this nature, we clearly cannot develop the full assemblage of tools that are a part of modern wavelet theory. [See [HERG, MEY1, MEY2, DAU] for more extensive treatments of this beautiful and dynamic subject. The papers [STR3] and [WAL] provide nice introductions as well.] What we can do is to give the reader a taste. Specifically, we shall develop a Multi-Resolution Analysis, or MRA; this study will show how Fourier analysis may be carried out with localization in either the space variable or the Fourier transform (frequency) variable. In short, the reader will see how either variable may be localized. Contrast this notion with the classical construction, in which the units are sines and cosines— clearly functions which *do not* have compact support. The exposition here derives from that in [HERG, STR3], and [WAL].

6.7.2 *Building a Custom Fourier Analysis*

Typical applications of classical Fourier analysis are to

- **Frequency Modulation:** Alternating current, radio transmission
- **Mathematics:** Ordinary and partial differential equations, analysis of linear and nonlinear operators

- **Medicine:** Electrocardiography, magnetic resonance imaging, biological neural systems
- **Optics and Fiber-Optic Communications:** Lens design, crystallography, image processing
- **Radio, Television, Music Recording:** Signal compression, signal reproduction, filtering
- **Spectral Analysis:** Identification of compounds in geology, chemistry, biochemistry, mass spectroscopy
- **Telecommunications:** Transmission and compression of signals, filtering of signals, frequency encoding

In fact, the applications of Fourier analysis are so pervasive that they are part of the very fabric of modern technological life.

The applications that are being developed for wavelet analysis are very similar to those just listed. But the wavelet algorithms give rise to faster and more accurate image compression, faster and more accurate signal compression, and better denoising techniques that preserve the original signal more completely. The applications in mathematics lead, in many situations, to better and more rapid convergence results.

What is lacking in classical Fourier analysis can be readily seen by examining the Dirac delta mass. Because, if the unit ball of L^1—thought of as a subspace of the dual space of $C(\mathbb{T})$—had any extremal functions (it does not), they would be objects of this sort: the weak-$*$ limit of functions of the form $N^{-1}\chi_{[-1/2N,1/2N]}$ as $N \to +\infty$. That weak-$*$ limit is the Dirac mass. We know the Dirac mass as the functional that assigns to each continuous function with compact support its value at 0:

$$\delta : C_c(\mathbb{R}^N) \ni \phi \longmapsto \phi(0).$$

It is most convenient to think of this functional as a measure:

$$\int \phi(x)\, d\delta(x) = \phi(0).$$

Now suppose that we want to understand δ by examining its Fourier transform. For simplicity, restrict attention to $\mathbb{R}^1$:

$$\widehat{\delta}(\xi) = \int_{\mathbb{R}} e^{i\xi \cdot t}\, d\delta(t) \equiv 1.$$

In other words, the Fourier transform of δ is the constant, identically 1, function. To recover δ from its Fourier transform, we would have to make sense of the inverse Fourier integral

$$\int 1 \cdot e^{-i\xi \cdot t}\, dt.$$

Doing so requires a careful examination of the Gauss–Weierstrass summation method and certainly strains the intuition: why should we have to "sum" exponentials, each of which is supported on the entire line and none of which is in any L^p class for $1 \leq p < \infty$, in order to re-construct δ—which is supported just at the origin?

The point comes through perhaps even more strikingly by way of Fourier series. Consider the Dirac mass δ supported at the origin in the circle group $\mathbb{T}$. Then the Fourier–Stieltjes coefficients of δ are

$$\widehat{\delta}(j) \equiv \frac{1}{2\pi} \int_{-\pi}^{\pi} e^{-ijt} \, d\delta(t) = 1.$$

Thus recovering δ from its Fourier series amounts to finding a way to sum the formal series

$$\sum_{j=-\infty}^{\infty} 1 \cdot e^{ijt}$$

in order to obtain the Dirac mass. Since each exponential is supported on the entire circle group, the imagination is defied to understand how these exponentials could sum to a point mass. [To be fair, the physicists have no trouble seeing this point: at the origin the terms all add up, and away from zero they all cancel out.]

The study of the point mass is not merely an affectation. In a radio signal, noise (in the form of spikes) is frequently a sum of point masses. On a phonograph record, the pops and clicks that come from imperfections in the surface of the record exhibit themselves (on an oscilloscope, for instance) as spikes, or point masses.

For the sake of contrast, in the next section we shall generate an *ad hoc* family of wavelet-like basis elements for L^2 and show how these may be used much more efficiently to decompose the Dirac mass into basis elements.

6.7.3 The Haar Basis

In this section we shall describe the Haar wavelet basis. While the basis elements are not smooth functions (as wavelet basis elements usually are—thanks to work of I. Daubechies [DAU]), they will exhibit the other important features of a Multi-Resolution Analysis (MRA). In fact we shall follow the axiomatic treatment as developed by Mallat and exposited in [WAL] in order to isolate the essential properties of an MRA.

We shall produce a dyadic version of the wavelet theory. Certainly other theories, based on other dilation paradigms, may be produced. But the dyadic theory is the most standard and quickly gives the flavor of the construction. In this discussion we shall use the notation α_δ to denote the dilate of a function: $\alpha_\delta f(x) \equiv f(\delta x)$. And we shall use the notation τ_a to denote the translate of a function: $\tau_a f(x) \equiv f(x - a)$.

We work on the real line $\mathbb{R}$. Our universe of functions will be $L^2(\mathbb{R})$. Define

$$\phi = \chi_{[0,1)}$$

and

$$\psi(x) \equiv \phi(2x) - \phi(2x - 1) = \chi_{[0,1/2)}(x) - \chi_{[1/2,1)}(x).$$

The reader should draw a figure so as to envision these functions.

The function ϕ will be called a *scaling function* and the function ψ will be called the associated *wavelet*. The basic idea is this: translates of ϕ will generate a space V_0 that can be used to analyze a function f on a large scale—more precisely, on the scale of size 1 (because 1 is the length of the support of ϕ). But the elements of the space V_0 cannot be used to detect information that is at a scale *smaller* than 1. So we will scale the elements of V_0 down by a factor of 2^j, each $j = 1, 2, \ldots,$ to obtain a space that can be used for analysis at the scale 2^{-j}. (and we will also scale V_0 *up* to obtain elements that are useful at an arbitrarily large scale.) Let us complete this program now for the specific ϕ that we have defined above and then present some axioms that will describe how this process can be performed in a fairly general setting.

Now we use ϕ to generate a scale of function spaces $\{V_j\}_{j \in \mathbb{Z}}$. We set

$$V_0 = \left\{ \sum_{k \in \mathbb{Z}} a_k[\tau_k \phi] : \sum |a_k|^2 < \infty \right\},$$

for the particular function ϕ that was specified above. Of course each element of V_0 so specified lies in L^2. (Because the functions $\tau_k \phi$ have disjoint supports.) But it would be wrong to think that V_0 is all of L^2, for an element of V_0 is constant on each interval $[k, k + 1)$, and has possible jump discontinuities only at the integers. The functions $\{\tau_k \phi\}_{k \in \mathbb{Z}}$ form an orthonormal basis (with respect to the L^2 inner product) for V_0.

Now let us say that a function g is in V_1 if and only if $\alpha_{1/2} g$ lies in V_0. Thus $g \in V_1$ means that g is constant on the intervals determined by the lattice $(1/2)\mathbb{Z} \equiv \{n/2 : n \in \mathbb{Z}\}$ and has possible jump discontinuities only at the elements of $(1/2)\mathbb{Z}$. It is easy to see that the functions $\{\sqrt{2}\alpha_2 \tau_k \phi\}$ form an orthonormal basis for V_1.

Observe that $V_0 \subseteq V_1$ since every jump point for elements of V_0 is also a jump point for elements of V_1 (but not conversely). More explicitly, we may write

$$\tau_k \phi = \alpha_2 \tau_{2k} \phi + \alpha_2 \tau_{2k+1} \phi,$$

thus expressing an element of V_0 as a linear combination of elements of V_1.

Now that we have the idea down, we may iterate it to define the spaces V_j for any $j \in \mathbb{Z}$. Namely, for $j \in \mathbb{Z}$, V_j will be generated by the functions $\alpha_{2^j}\tau_m\phi$, all $m \in \mathbb{Z}$. In fact we may see explicitly that an element of V_j will be a function of the form

$$f = \sum_{\ell \in \mathbb{Z}} a_\ell \chi_{[\ell/2^j,[\ell+1]/2^j)},$$

where $\sum |a_\ell|^2 < \infty$. Thus an orthonormal basis for V_j is given by $\{2^{j/2}\alpha_{2^j}\tau_m\phi\}_{m\in\mathbb{Z}}$.

Now the spaces V_j have no common intersection except the zero function. This is so because, since a function $f \in \cap_{j\in\mathbb{Z}} V_j$ would be constant on arbitrarily large intervals (of length 2^{-j} for j negative), then it can only be in L^2 if it is zero. Also $\cup_{j\in\mathbb{Z}} V_j$ is dense in L^2 because any L^2 function can be approximated by a simple function (i.e., a finite linear combination of characteristic functions), and any characteristic function can be approximated by a sum of characteristic functions of dyadic intervals.

We, therefore, might suspect that if we combine all the orthonormal bases for all the V_j, $j \in \mathbb{Z}$, then this would give an orthonormal basis for L^2. That supposition is, however, incorrect. For the basis elements $\phi \in V_0$ and $\alpha_{2^j}\tau_0\phi \in V_j$ are not orthogonal. This is where the function ψ comes in.

Since $V_0 \subseteq V_1$ we may proceed by trying to complete the orthonormal basis $\{\tau_k\phi\}$ of V_0 to an orthonormal basis for V_1. Put in other words, we write $V_1 \equiv V_0 \oplus W_0$, and we endeavor to write a basis for W_0. Let $\psi = \alpha_2\phi - \alpha_2\tau_1\phi$ be as above, and consider the set of functions $\{\tau_m\psi\}$. Then this is an orthonormal set. Let us see that it spans W_0.

Let h be an arbitrary element of W_0. So certainly $h \in V_1$. It follows that

$$h = \sum_j b_j a_2 \tau_j \phi$$

for some constants $\{b_j\}$ that are square-summable. Of course h is constant on the interval $[0, 1/2)$ and also constant on the interval $[1/2, 1)$. We note that

$$\phi(t) = \frac{1}{2}[\phi(t) + \psi(t)] \qquad \text{on } [0, 1/2)$$

and

$$\phi(t) = \frac{1}{2}[\phi(t) - \psi(t)] \qquad \text{on } [1/2, 1).$$

It follows that

$$h(t) = \left(\frac{b_0 + b_1}{2}\right)\phi(t) + \left(\frac{b_0 - b_1}{2}\right)\psi(t)$$

on $[0, 1)$. Of course a similar decomposition obtains on every interval $[j, j + 1)$.
   As a result,

$$h = \sum_{j \in \mathbb{Z}} c_j \tau_j \phi + \sum_{j \in \mathbb{Z}} d_j \tau_j \psi,$$

where

$$c_j = \frac{b_j + b_{j+1}}{2} \quad \text{and} \quad d_j = \frac{b_j - b_{j+1}}{2}.$$

Note that $h \in W_0$ implies that $h \in V_0^{\perp}$. Also every $\tau_j \phi$ is orthogonal to every $\tau_k \psi$. Consequently every coefficient $c_j = 0$. Thus we have proved that h is in the closed span of the terms $\tau_j \psi$. In other words, the functions $\{\tau_j \psi\}_{j \in \mathbb{Z}}$ span W_0.
   Thus we have $V_1 = V_0 \oplus W_0$, and we have an explicit orthonormal basis for W_0. Of course we may scale this construction up and down to obtain

$$V_{j+1} = V_j \oplus W_j \tag{6.1}$$

for every j. And we have the explicit orthonormal basis $\{2^{j/2} \alpha_{2^j} \tau_m \psi\}_{m \in \mathbb{Z}}$ for each W_j.
   We may iterate Eq. $(6.1)_j$ to obtain

$$V_{j+1} = V_j \oplus W_j = V_{j-1} \oplus W_{j-1} \oplus W_j$$
$$= \cdots = V_0 \oplus W_0 \oplus W_1 \oplus \cdots \oplus W_{j-1} \oplus W_j.$$

Letting $j \to +\infty$ yields

$$L^2 = V_0 \oplus \bigoplus_{j=0}^{\infty} W_j. \tag{6.2}$$

But a similar decomposition may be performed on V_0, with W_j in descending order:

$$V_0 = V_{-1} \oplus W_{-1} = \cdots = V_{-\ell} \oplus W_{-\ell} \oplus \cdots \oplus W_{-1}.$$

Letting $\ell \to +\infty$, and substituting the result into (6.2), now yields that

$$L^2 = \bigoplus_{j \in \mathbb{Z}} W_j.$$

Thus we have decomposed $L^2(\mathbb{R})$ as an orthonormal sum of Haar wavelet subspaces. We formulate one of our main conclusions as a theorem:

Theorem 6.7.1 *The collection*

$$\mathcal{H} \equiv \left\{ \alpha_{2^j} \tau_m \psi : m, j \in \mathbb{Z} \right\}$$

is an orthonormal basis for L^2 and will be called a wavelet basis for L^2.

Now it is time to axiomatize the construction that we have just performed in a special instance.

Axioms for a Multi-Resolution Analysis (MRA)

A collection of subspaces $\{V_j\}_{j \in \mathbb{Z}}$ of $L^2(\mathbb{R})$ is called a *Multi-Resolution Analysis* or MRA (this is an idea of Mallat [MAL]) if

MRA_1 **(Scaling)** For each j, the function $f \in V_j$ if and only if $\alpha_2 f \in V_{j+1}$.
MRA_2 **(Inclusion)** For each j, $V_j \subseteq V_{j+1}$.
MRA_3 **(Density)** The union of the V_j's is dense in L^2:

$$\text{closure} \left\{ \bigcup_{j \in \mathbb{Z}} V_j \right\} = L^2(\mathbb{R}).$$

MRA_4 **(Maximality)** The spaces V_j have no nontrivial common intersection:

$$\bigcap_{j \in \mathbb{Z}} V_j = \{0\}.$$

MRA_5 **(Basis)** There is a function ϕ such that $\{\tau_j \phi\}_{j \in \mathbb{Z}}$ is an orthonormal basis for V_0.

We invite the reader to review our discussion of $\phi = \chi_{[0,1)}$ and its dilates and confirm that the spaces V_j that we constructed do indeed form an MRA. Notice in particular that, once the space V_0 has been defined, then the other V_j are completely and uniquely determined by the MRA axioms.

Chapter 7
Concluding Remarks

Certainly the real variable theory of Hardy spaces has been one of the dominant ideas in the harmonic analysis of the second half of the twentieth century. E. M. Stein played the pivotal role in the development of these ideas, beginning with his 1960 paper with Guido Weiss, and continuing with the seminal 1972 paper with Charles Fefferman. Through his many talented Ph.D. students, the influence and permeation of Stein's ideas has been considerable.

The purpose of this volume is to pay tribute to Stein's insights, and to contribute to his memory. I am sure that there are hundreds of others who share this author's feelings.

© The Author(s), under exclusive license to Springer Nature Switzerland AG 2023   243
S. G. Krantz, *The E. M. Stein Lectures on Hardy Spaces*, Lecture Notes in
Mathematics 2326, https://doi.org/10.1007/978-3-031-21952-8_7

References

[BGS] D. Burkholder, R. Gundy, M. Silverstein, A maximal function characterization of the class H^p. Trans. Amer. Math. Soc. **157**, 137–153 (1971)

[CAL] A.P. Calderón, On the behaviour of harmonic functions at the boundary. Trans. Amer. Math. Soc. **68**, 47–54 (1950)

[CAT] A.P. Calderón, A. Torchinsky, Parabolic maximal functions associated with a distribution. Adv. Math. **16**, 1–64 (1975)

[CAZ1] A.P. Calderón, A. Zygmund, On the existence of certain singular integrals. Acta Math. **88**, 85–139 (1952)

[CAZ2] A.P. Calderón, A. Zygmund, Algebras of certain singular operators. Amer. J. Math **78**, 310–320 (1956)

[CAR] L. Carleson, A counterexample for measures bounded on H^p spaces for the bidisk, Mittag-Leffler Report No. 7, Institute Mittag-Leffler, Djursholm (1974)

[CHF] S.Y. Chang, R. Fefferman, A continuous version of duality of H^1 with BMO on the bidisc. Ann. Math. **112**, 179–201 (1980)

[CHG] M. Christ, D. Geller, Singular integral characterizations of Hardy spaces on homogeneous groups. Duke Math. J. **51**, 547–598 (1984)

[CKS1] D.-C. Chang, S.G. Krantz, E.M. Stein, H^p theory on a smooth domain in $\mathbb{R}^N$ and elliptic boundary value problems. J. Funct. Anal. **114**, 286–347 (1993)

[CKS2] D.-C. Chang, S.G. Krantz, E.M. Stein, Hardy spaces and elliptic boundary value problems, in *The Madison Symposium on Complex Analysis* (Madison, WI, 1991), pp 119–131, Contemporary Mathematics, vol. 137 (American Mathematical Society, Providence, 1992)

[COI] R.R. Coifman, A real variable characterization of H^p. Studia Math. **51**, 269–274 (1974)

[COW] R.R. Coifman, G. Weiss, Extensions of Hardy spaces and their use in analysis. Bull. Amer. Math. Soc. **83**, 569–645 (1977)

[CRTW] R.R. Coifman, R. Rochberg, M. Taibleson, G. Weiss, *Representation Theorems for Hardy Spaces*. Astérisque, vol. 77, Society Mathematical France, Paris (Cambridge University Press, Cambridge, 1980), pp. 1–9

[DAH] B. Dahlberg, Estimates of harmonic measure. Arch. Rational Mech. Anal. **65**, 275–288 (1977)

[DAJ] G. David, J.-L. Journé, A boundedness criterion for generalized Calderó–Zygmund operators. Ann. Math. **120**, 371–397 (1984)

[DAU] I. Daubechies, *Ten Lectures on Wavelets* (Society for Industrial and Applied Mathematics, Philadelphia, 1992)

[DUR] P. Duren, *Theory of H^p Spaces* (Academic, New York, 1970)

© The Author(s), under exclusive license to Springer Nature Switzerland AG 2023   245
S. G. Krantz, *The E. M. Stein Lectures on Hardy Spaces*, Lecture Notes in
Mathematics 2326, https://doi.org/10.1007/978-3-031-21952-8

[FEF1] C. Fefferman, The multiplier problem for the ball. Ann. Math. **94**, 330–336 (1971)

[FEP] R. Fefferman and J. Pipher, Multiparameter operators and sharp weighted inequalities. Amer. J. Math. **119**, 337–369 (1997)

[FES] C. Fefferman and E. M. Stein, H^p spaces of several variables. Acta Math. **129**, 137–193 (1972)

[FOS] G.B. Folland, E.M. Stein, Estimates for the $\bar{\partial}_b$ complex and analysis on the Heisenberg group. Comm. Pure Appl. Math. **27**, 429–522 (1974)

[FRJ] M. Frazier, B. Jawerth, A discrete transform and decompositions of distribution spaces. J. Funct. Anal. **93**, 34–170 (1990)

[GOL] D. Goldberg, A local version of real Hardy spaces. Duke Math. J. **46**, 27–42 (1979)

[HER] C. Herz, Bounded mean oscillation and regulated martingales. Trans. Amer. Math. Soc. **193**, 199–215 (1974)

[HERG] E. Hernández, G. Weiss, *A First Course on Wavelets*. With a Foreword by Yves Meyer. Studies in Advanced Mathematics (CRC Press, Boca Raton, 1996)

[HOF] K. Hoffman, *Banach Spaces of Analytic Functions* (Prentice-Hall, Englewood Cliffs, 1962)

[JON] F. John, L. Nirenberg, On functions of bounded mean oscillation. Comm. Pure Appl. Math. **14**, 415–426 (1961)

[JOUR1] J.-L. Journé, Two problems of Calderón–Zygmund theory on product-spaces. Ann. Inst. Fourier (Grenoble) **38**, 111–132 (1988)

[JOUR2] J.-L. Journé, Calderón-Zygmund operators on product spaces. Rev. Mat. Iberoamericana **1**, 55–91 (1985)

[KOO] P. Koosis, *Introduction to H_p Spaces*. With an appendix on Wolff's proof of the corona theorem, London Mathematical Society Lecture Note Series, vol. 40 (Cambridge University Press, Cambridge, 1980)

[KRA1] S.G. Krantz, *Explorations in Harmonic Analysis, with Applications in Complex Function Theory and the Heisenberg Group* (Birkhäuser Publishing, Boston, 2009)

[KRA2] S.G. Krantz, *Geometric Analysis and Function Spaces* (CBMS and the American Mathematical Society, Providence, 1993)

[KRA3] S.G. Krantz, A new theory of atomic H^p spaces with applications to smoothness of functions, preprint

[KRA4] S.G. Krantz, *Function Theory of Several Complex Variables*, 2nd edn. (American Mathematical Society, Providence, 2001)

[KRA5] S.G. Krantz, Lipschitz spaces, smoothness of functions, and approximation theory. Exposition. Math. **1**, 193–260 (1983)

[KRA6] S. Krantz, *Partial Differential Equations and Complex Analysis* (CRC Press, Boca Raton, 1992)

[LAT] R. Latter, A characterization of $H^p(\mathbb{R}^n)$ in terms of atoms. Studia Math. **62**, 93–101 (1978)

[LWW] D. Labate, G. Weiss, E. Wilson, Wavelets. Notices Amer. Math. Soc. **60**, 66–76 (2013)

[MAL] S. Mallat, *A Wavelet Tour of Signal Processing*. The Sparse Way, 3rd edn. With Contributions from Gabriel Peyré (Elsevier/Academic, Amsterdam, 2009)

[MER] K. Merryfield, On the area integral, Carleson measures and H^p in the polydisc. Indiana Univ. Math. J. **34**, 663–685 (1985)

[MEY1] Y. Meyer, *Wavelets and Operators*, Translated from the 1990 French original by D. H. Salinger, Cambridge Studies in Advanced Mathematics, vol. 37 (Cambridge University Press, Cambridge, 1992)

[MEY2] Y. Meyer, *Wavelets: Algorithms and Applications*, Translated from the original French and with a Forward by Robert D. Ryan (SIAM, Philadelphia, 1993)

[RFEF] R. Fefferman, Bounded mean oscillation on the polydisk. Ann. of Math. **110**, 395–406 (1979)

[RFES] R. Fefferman and E. M. Stein, Singular integrals on product spaces. Adv. in Math. **45**, 117–143 (1982).

[RUD1] W. Rudin, *Function Theory in Polydiscs* (W. A. Benjamin, New York, 1969)

[RUD2] W. Rudin, *Real and Complex Analysis* (McGraw-Hill, New York, 1966)

[SAW] S. Sawyer, Maximal inequalities of weak type. Ann. Math. **84**, 157–174 (1966)

[STE1] E.M. Stein, *Singular Integrals and Differentiability Properties of Functions* (Princeton University Press, Princeton, 1970)

[STE2] E.M. Stein, *Boundary Behavior of Holomorphic Functions of Several Complex Variables* (Princeton University Press, Princeton, 1972)

[STE3] E.M. Stein, The analogues of Fatou's theorem and estimates for maximal functions, in *1968 Geometry of Homogeneous Bounded Domains* (C.I.M.E., 3 Ciclo, Urbino, 1967), Edizioni Cremonese, Rome

[STE4] E.M. Stein, On limits of sequences of operators. Ann. Math. **74**, 140–170 (1961)

[STR3] R. Strichartz, How to make wavelets. Am. Math. Monthly **100**, 539–556 (1993)

[STN] E.M. Stein, N.J. Weiss, On the convergence of Poisson integrals. Trans. AMS **140**, 35–54 (1969)

[STW1] E.M. Stein, G. Weiss, *Introduction to Fourier Analysis on Euclidean Spaces* (Princeton University Press, Princeton, 1971)

[STW2] E.M. Stein, G. Weiss, On the theory of harmonic functions of several variables. I. The theory of H^p-spaces. Acta Math. **103**, 25–62 (1960)

[TAW] M. Taibleson, G. Weiss, The molecular characterization of certain Hardy spaces, in *Representation Theorems for Hardy Spaces*. Astérisque, vol. 77, Society Mathematical France, Paris, (Cambridge University Press, Cambridge, 1980), pp. 67–149

[WAL] J.S. Walker, Fourier analysis and wavelet analysis. Notices AMS **44**, 658–670 (1997)

[WID] K.O. Widman, On the boundary values of harmonic functions in $\mathbb{R}^3$. Ark. Mat. **5**, 221–230 (1964)

[WOL1] T. Wolff, A Kakeya-type problem for circles. Amer. J. Math. **119**, 985–1026 (1997)

[WOL2] T. Wolff, An improved bound for Kakeya type maximal functions. Rev. Mat. Iberoamericana **11**, 651–674 (1995)

[WOL3] T. Wolff, Counterexamples with harmonic gradients in $\mathbb{R}^3$, in *Essays on Fourier Analysis in Honor of Elias M. Stein* (Princeton, NJ, 1991), pp. 321–384; Princeton Mathematical Series, vol. 42 (Princeton Univ. Press, Princeton, 1995)

[ZYG] A. Zygmund, *Trigonometric Series*, vols. I, II, 2nd edn. (Cambridge University Press, London, 1968)

Index

© The Author(s), under exclusive license to Springer Nature Switzerland AG 2023
S. G. Krantz, *The E. M. Stein Lectures on Hardy Spaces*, Lecture Notes in
Mathematics 2326, https://doi.org/10.1007/978-3-031-21952-8

Printed in the United States
by Baker & Taylor Publisher Services

Printed in the United States
by Baker & Taylor Publisher Services